Practical
Techniques
for
GROUNDWATER
and
SOIL
REMEDIATION

Evan K. Nyer

LEWIS PUBLISHERS
Boca Raton   Ann Arbor   London   Tokyo

Library of Congress Cataloging-in-Publication Data

Nyer, Evan K.
  Practical techniques for groundwater and soil remediation
/ by Evan K. Nyer.
      p.   cm. -- (Geraghty & Miller science & engineering series)
  Includes bibliographical references and index.
  ISBN 0-87371-731-7
  1. Water, Underground--Purification.   2. Soil pollution.
3. Bioremediation.   I. Title.   II. Series.
TD426.N95   1992
628.1'68--dc20   92-9270
                                                                   CIP

ISBN 0-87371-731-7

Direct all inquiries to CRC Press, Inc., 2000 Corporate Blvd., N.W., Boca
Raton, FL 33431

LEWIS PUBLISHERS
121 South Main Street, Chelsea, Michigan 48118

PRINTED IN MEXICO   1 2 3 4 5 6 7 8 9 0

Printed on acid-free paper

Series Preface

If it is true that corporations, like individuals, have well-defined personalities, then Geraghty & Miller, Inc. must surely be thought of as a straightforward and practical organization. It was Brian Lewis, of Lewis Publishers, who, from a publisher's perspective, noticed this characteristic and the growing concentration of talented environmental professionals assembling within G & M. Brian also saw practical books — handbooks, field guides, and desktop references — books designed to be used, coming from authors who were currently and deeply involved in solving environmental problems.

Organizing a full series of books to be authored within the structure of one company presents unusual problems. Publishing schedules have to be made entirely flexible as to not interfere with regular project work, and the publisher, editor, and authors must each sustain interest through many interruptions. Fortunately, Geraghty & Miller is no stranger to the publishing process, having established its own in-house publishing subsidiary, Water Information Center, Inc., over 30 years ago. This subsidiary, unique in the environmental services industry, provides the extra, internal support necessary to keep the series organized. Although primarily produced by Lewis Publishers, the series will contain some books that bear imprints of both publishers, signaling a co-publishing agreement on that specific title.

At its inception, this series was identified as a "Groundwater Series" since that subject has been a speciality of G & M for 35 years. But when the first book, *The Water Encyclopedia,* came along, it became clear that the series would reflect the current full-service nature of Geraghty & Miller, Inc. and span all environmental science and engineering topics. The format size and production method used for each book in the series will also vary considerably with the planned use of the book.

In both subject matter and design, the series will try to provide very practical books that are meant to be among the most frequently used in your library.

Fred L. Troise
Series Editor
Vice President/Marketing, Geraghty & Miller, Inc.
President, Water Information Center, Inc.

About the Author

Evan K. Nyer is an expert in the research and application of technology to groundwater cleanups. As Vice President with Geraghty & Miller, Inc., he is responsible for engineering services including hazardous and solid waste management, environmental and natural resource management, remediation activities, and designing treatment systems for contaminated sites throughout the United States and in foreign countries. He has designed over 100 groundwater treatment systems.

Mr. Nyer travels throughout the country teaching treatment techniques at seminars and universities. He has written numerous papers on groundwater decontamination and other water and wastewater cleanup techniques. He is responsible for bringing to the field many new innovative techniques for biological treatment of water, soils, and in situ treatment and the application of existing technologies to groundwater contamination. He is a member of the Water Pollution Control Federation, National Water Well Association, American Institute of Chemical Engineers, and the American Society of Civil Engineers.

With love, to my parents,
   Alice and Nealan Nyer

Contents

Section VI
Selection of Treatment Alternatives

Section VII
Practical Problems

Section VIII
In Situ and Natural Biochemical Remediations

INTRODUCTION

Since 1987, the National Ground Water Association (NGWA), formerly the National Water Well Association (NWWA), has provided space in their magazine, *Groundwater Monitoring Review*, for me to write about groundwater and soil remediations. The magazine is published four times each year and my section is titled "Treatment Technology". This book is a summary of the articles from 1987 through the spring of 1992.

I would like to thank the NGWA for their help and cooperation over the past six years. The association has given me wide latitude to express a variety of viewpoints with my articles. Even though some articles proved very controversial, they were neither altered nor was I prompted to modify their original intent to tone them down. This artistic freedom has allowed me to provide information that is not usually published (treatment systems that did not work in the field), and to write in a form that was conversational and easy to read. I hope that readers have found the articles interesting over the years and that they appreciate the daring and foresight that the NGWA has promoted with this writing style. Special thanks are given to Anita Stanley, the Managing Editor of *Groundwater Monitoring Review*, and Jay Lehr, the former Executive Director of NGWA.

I have placed all of the magazine articles into a book for several reasons. First, some of the articles are good reference sources. Detailed data is provided for specific compounds and treatment sources. Providing this data in one location makes the book a very useful reference source. One area, for example, that should be very helpful is in Section II, Chapter 3, "Using the Properties

ISBN 0-87371-731-7
© 1992 by Lewis Publishers

1

of Organic Compounds to Help Design a Treatment System''. The solubility, specific gravity, octanol/water partition coefficient, Henry's Law constant, carbon adsorption capacity, and the biodegradability of fifty organic compounds is described in this chapter. These fifty compounds are the most common constituents reported at contaminated sites. Other articles contain useful reference material.

The second reason for the book is the practical nature of the articles. Most technical papers provide scientific information and field data. These papers do not cover the practical aspects of remediations. There are very few papers based upon failures. There are few papers based on field experiences. The format of the ''Treatment Technology'' section has allowed me to write about the mundane aspects of the remediation. These design and operational details of a project are very important. This book emphasizes the practical aspects of a remediation by putting all of these papers in one place. The four chapters in Section VII are dedicated to practical problems.

The third reason for organizing the book is experience. The groundwater consulting industry has progressed rapidly since the early 1980s. Less than ten years ago, when we envisioned a contaminated site, we pictured scarred land and leaking 55-gallon drums, stacked four or five high or strewn carelessly across a site. A perfect example of this is the Seymour superfund site in Seymour, Indiana. The initial photographs of the site show an immense collection of 55-gallon drums, large storage tanks, stained soil, and other obvious signs of neglect. We have been working on that site for more than eight years now. The barrels have been removed; the contaminated soil has been contained by a RCRA cap and vapors are being treated by a Vapor Extraction System (VES); the ground water is being captured by a series of interception wells and recovered ground water is being treated by air stripping and carbon adsorption treatment systems. The groundwater treatment system has been operational for over four years. My history with this system and that of over 100 other groundwater and soil treatment systems has provided a wealth of personal experience.

Similar systems now have provided in-depth experience with operating groundwater treatment systems. We no longer have to guess. We can see the actual results and compare them to our original ideas on what would happen as we conducted the remediation program. Chapter 18 in Section VIII titled ''Biochemical Effects on Contaminants' Fate and Transport'' provides actual data on an operating Seymour groundwater system versus predicted results.

The fourth and final reason for the book is strategy. We now know enough about groundwater and soil remediation that we can develop a strategy at the beginning of the project. Too often I see managers approaching remediation as only a step-by-step process. The only way to get to the next step is to perform the previous step. This is no longer true. We know the general pattern that the remediation will take. The details will always be different and have to be investigated. If we do not have a project strategy from the beginning we will end up wasting both time and money. The book will help the reader prepare a strategy before the remediation project starts.

While the topics listed are important specific subjects, the book is organized to be read either as an entire volume or for the reader to go directly to the section or chapter that is of main interest. Although the chapters are easy to read, the book also includes detailed technical data on the treatment equipment performance and costs associated with their design and operation. The book is unique in that it also includes data from treatment systems that did not work. Section V provides specific information on the operations of groundwater treatment equipment.

Finally, I would like to thank the other authors for their contributions to the original articles and, now, to this book. Over the years I have requested the help from various professionals on specific articles. Collectively they have spent many hours helping me gather specific technical information and putting it into a form that would be useful to the reader. When their efforts included writing, they were made co-authors of that article. The book has listed the authors of each chapter so that their contribution is recognized. I can easily say that some of the best work that I have done has been in conjunction with these individuals. One other person that must be thanked is Ralph Moon. Ralph has the bad luck of having good technical sense and writing ability. For the past couple of years he has been reviewing the final draft of the articles. His input has been greatly appreciated.

I hope that the readers find the material useful and that they get the main point of the book which is that most of the time a good, practical design is more important than the latest, most advanced technology.

SECTION I

Starting the Project

This section of the book is the first paper that I wrote for the column in *Groundwater Monitoring Review*. This book is not presented in the chronological order of the original publication of the papers. However, I feel strongly that "Defining Treatment Parameters" should be the first subject that is covered. It is also the first subject that is covered in my textbook, *Groundwater Treatment Technology*.

When I went to college we were taught to begin an engineering project by drawing a black box for the treatment system. We then drew arrows entering and leaving the box that represented the influent, required effluent, and any byproducts produced by the treatment process. By defining these arrows we, in effect, defined the treatment system.

The black box should be the first step of any groundwater or soil remediation design. Some of the process data will not be available at the beginning of the project; however, this lack of data also helps to define the design. We now know the information that must be gathered during our investigation in order to complete the treatment system design. Too often we assume that the data collected during the investigation will be sufficient for the treatment design. To approach the project correctly we must have some idea of the treatment system before the investigation is finished to make sure that we get all of the required information.

Data must not be limited to the influent conditions. We must also understand the effluent requirements and any byproducts that may be produced. Solids generation and air emissions, for example, can be an important part of selecting one design over another. The basic point is that the designer must have some idea of the treatment objectives before starting off in a particular direction.

CHAPTER 1

Defining Treatment Parameters*

Evan K. Nyer

Treatment of contaminants in ground water is a relatively new field. While it is closely related to waste water treatment, the specific design requirements are unique. We are presently installing treatment systems based on theory and pilot plant only. We are not yet able to rely on years of experience from operating treatment systems for design specifications.

New methods of applying existing technology, and new technologies themselves, are being placed in the field at a rapid rate. Future installations will not be able to wait until these new methods are installed, evaluated, and the results published.

The best treatment technology depends upon more than just the contaminant. It will also depend upon the concentration, flow, government regula-

* This is the first "Treatment Technology" column which appeared in Fall 1987 issue of *Groundwater Monitoring Review*.

ISBN 0-87371-731-7
© 1992 by Lewis Publishers

tions, etc. The more information that we have on the contaminated ground water, the more accurate and complete the response can be.

We need to define the treatment parameters before we can proceed with selecting the optimum treatment technology.

TREATMENT PARAMETERS

The first step in any aquifer cleanup is to define the treatment parameters that will be used to design the treatment system. The most important design parameters are: flow, concentration, and effluent requirements.

Flow

The first design criteria that we need to consider is the flow to the aboveground treatment system. Because we will not be doing detailed designs, we do not need exact numbers. However, one treatment technology may be applicable at 10 gpm and a different technology applicable at 100 gpm. Flow rates should be provided even for cases in which flow will not be a design parameter. For example, in situ methods do not relate directly to flow, but we will want to compare aboveground methods to the in situ methods. We will need the flow rates for this comparison.

Concentration and Quantity

The first piece of information that is required is a list of the compounds that have to be removed from the ground water. In a basic, hypothetical case history it would not be practical to discuss a ground water with 50 to 100 compounds in it. A single compound is preferred, with a limitation of five compounds. In a case history with more than five compounds, we would try to analyze the alternatives based upon the compounds that will control the design.

Next, we will need the concentration and quantity of the contaminant(s) to be treated. The concentration of inorganic and organic material should be reported in ppm or ppb. The quantity (if available) of the contaminant should be given in pounds.

We may also need information not limited to the specific contaminant found in the ground water. Successful treatment of inorganic contaminants will also require a knowledge of other divalent cations present in the water. Either a hardness measurement or specific calcium and magnesium concentrations are needed. Divalent cations will be affected by most methods used to remove inorganic material.

Treatment of organic contaminants also requires information about other material present in the ground water. For example, it is always a good idea to analyze for general total organic content when dealing with organic contaminants. Simply summing the concentration of specific organics does not guarantee a knowledge of the total organics present. Gas chromatography/ mass spectroscopy cannot identify every organic compound in existence. Unknown compounds released to the environment and transformation of the known compounds can produce a background of unknown organic compounds. The performance of a treatment system may depend on the total concentration of organics contained in the ground water. We suggest the use of Total Organic Carbon (TOC) as a measure of general organic content.

The quantity of the material present is also important to the final design. The total mass affects the design in two important ways. First, the total length of time needed for a cleanup will be related to the total mass of material released to the environment. Second, the concentration of contaminant coming from the well can change over time. Sometimes the amount of material released is not known. In those cases, one should estimate the design effects directly.

The first design effect is: how long will the cleanup take? Certain technology is preferred for short-term projects. Emergency conditions usually require carbon adsorption for organic material. Longer term projects can consider the different advantages between carbon, air stripping, and biological treatment. We need to know whether the project is going to take three months or three years.

Will the concentration change over time? If the source of the contaminant has not been removed (landfill leachate, for example), then the concentration will remain the same. If the source of contamination has been removed (leaking underground storage tank for example), then the concentration will decrease during the cleanup. A change in concentration will definitely affect the design of the treatment system. We do not need the rate of change, only the fact that it will change.

Effluent Criteria

We will need information on where the ground water will be released and the required concentrations of the contaminant for discharge. Some technologies are not appropriate for certain water usages. For example, we would not recommend biological treatment for water that was to be used as drinking water.

In theory, getting the required effluent concentrations should be a straightforward process. The effluent concentrations will affect the type of technology (or combinations of technologies) that can be applied to the ground water.

These levels are usually determined by local, state, and federal regulatory organizations. In practice, however, getting the regulatory agencies to set reasonable levels (levels that are technically based) can be quite difficult. Effluent levels of ''nondetectable'' or ''zero'' would not be considered reasonable, and we will assume all such values to be less than 1 ppb or the detection limit of that particular compound.

Regulations

We would need as much information about local and state regulations as possible. Some states do not allow in situ treatment. Some states do not allow air discharges from an air stripper. Some states have earthquake codes that add to construction costs of certain types of tanks and vessels.

These are just a few examples. The more information that we have on the regulations, the more accurate an analysis can be.

SUMMARY

Every groundwater treatment analysis should begin by defining the treatment parameters. In order to properly analyze the treatment alternatives the following basic information must be considered:

- list of contaminants to be removed
- flow to aboveground treatment system
- concentration of contaminants
- quantity of contaminants
- effluent requirements
- regulations controlling the installation.

SECTION II

Properties of the Contaminants

There are two important uses for the chemical properties of organic compounds at a remediation site. The first use is as part of the investigation. The properties of the compounds show the interaction of the organics with the environment around the organics. In fact, the only way that we understand what the data means is by interpreting the concentration of a compound in relation to the properties of that compound.

The best example of this use is retardation factors. We know that organic compounds do not migrate at the same rate as the free water in the aquifier. The chemical interaction of the organic compounds with the soil particles of the aquifer cause the organics to be retarded as they move through the aquifer. Retardation is a fundamental factor of contaminant treatment design. All remediations consider this chemical property when dealing with plume movement or recovery.

Other examples of the use of the chemical properties in investigations include the adsorption of compounds on soil and the volatility of compounds in the vadose zone. Adsorption can be used to determine the amount of material that is retained by the soil in the vadose zone and never reaches the aquifer. Volatility can be used to explain contaminant concentration measured in the vadose zone gas phase in relation to groundwater and soil concentrations.

A second important use of the chemical properties is in the design of the treatment systems that are used to remove the compounds from ground water and soils. Specific data is readily available on the treatability properties of each compound. This data can be used to prepare a preliminary analysis, preliminary designs, and cost estimates on a treatment method for a site.

An example of this use is the Henry's Law constant for the design of air strippers. This section lists the Henry's Law constant for fifty organic compounds. Based upon this data, the designer can estimate the removal efficiency that an air stripper will have on a particular compound. Using this chemical property, the designer can determine whether an air stripper should be studied further or eliminated as a possible treatment method. Carbon adsorption and biological treatment can be evaluated by the same method. This is a very cost effective method to evaluate treatment systems. Based upon chemical properties, most treatment methods can be selectively eliminated before significant amounts of money have been spent on the feasibility study.

This section provides the information necessary for these evaluations. The properties of most of the organic compounds that will be found at contaminated sites are included in these chapters. The first discusses chemical mixtures like gasoline. An important point to remember when dealing with a spill of a material like gasoline is that it is made up of several compounds. Each compound has its own properties. Any remediation design has to encompass all of the compounds (and their properties) if it is to be successful. The remaining chapters give the detailed information on the chemical properties and discusses how to utilize them in a project.

CHAPTER 2

Relating the Physical and Chemical Properties of Petroleum Hydrocarbons to Soil and Aquifer Remediation

Evan K. Nyer and George J. Skladany

This chapter is devoted to a detailed review of the physical and chemical properties of petroleum hydrocarbons. This review can then be related to different remediation methods. We have tried to provide sufficient data for the reader to use as reference material in future work.

Petroleum hydrocarbons are one of the most frequent sources of ground-water contamination. Leaking underground storage tanks at gasoline stations, spills at oil terminals, and leaks in fuel pipelines combine to make these compounds a recurring problem. Gasoline, diesel, and fuel oils are some of the most common petroleum products contaminating soils and ground water. While these products are generally spoken of as single entities, each is actually a complex mixture of many organic chemicals. Each of these specific

ISBN 0-87371-731-7
© 1992 by Lewis Publishers

chemicals has its own properties and behavior when in contact with soils and water. While it is correct to say that an aquifer has been polluted with the general contaminant "gasoline", remediation efforts must address the treatment of the specific organics present.

PHYSICAL AND CHEMICAL PROPERTIES

Chemical Composition

Let us start with simple definitions for gasoline, diesel, and fuel oils:[2]

- Gasoline is a mixture of volatile hydrocarbons suitable for use in internal combustion engines. The major chemical components of gasoline are branched chain paraffins (branched chain alkanes) cycloparaffins (cycloalkanes), and aromatics.
- Diesel is Number 2 fuel oil, composed primarily of unbranched paraffins (straight chain alkanes) with a flash point between 110° and 190°F (43° and 88°C).
- Fuel oils are chemical mixtures having flash points greater than 100°F (38°C). Fuel oils can be distilled fractions of petroleum, residuum from refinery operations, crude petroleum, or a mixture of two or more of these materials.

Basically, these different petroleum mixtures represent progressive "cuts" of a distillation column. Table 1 presents some of the major commercial products associated with different distillation fractions. Figure 1 shows some of the major petroleum hydrocarbon constituents as they would appear in a gas chromatograph separating compounds by increasing the boiling point. Gasoline is, in general, a mixture of chemicals with boiling points less than that of decane (those compounds with boiling points between 36° and 173°C). Gasoline contains relatively large concentrations of benzene, toluene, and xylene. Diesel fuels, on the other hand, consist primarily of higher boiling-point, straight chain alkanes. Diesel fuel contaminated soils therefore would not be expected to contain high concentrations of aromatic compounds.[5]

The source of the crude oil used for refining also has an effect on the composition of the final petroleum product. For example, Table 2 shows the volume percent of paraffins, cycloparaffins, and aromatics present in nine types of crude oil. Variability is again observed in the gasoline fraction produced from three crude oils, as shown in Table 3. For example, the gasoline fraction made from Conroe, Texas crude oil contains 3.27% benzene and 16.19% toluene on a volume basis. The gasoline fraction made from Colinga, California crude oil contains only 2.22% benzene and 7.94% toluene on a volume basis.[4]

Table 1. Petroleum Distillation Products

Fraction	Distillation Temperature, °C	Carbon Number
Gas	Below 20	C-1 to C-4
Petroleum ether	20 to 60	C-5 to C-6
Ligroin (light naphtha)	60 to 100	C-6 and C-7
Natural gasoline	40 to 205	C-5 to C-10 and cycloalkanes
Kerosene	175 to 325	C-12 to C-18 and aromatics
Gas oil	Above 275	C-12 and higher
Lubricating oil	Nonvolatile liquids	Probably long chains attached to cyclic compounds
Asphalt or petroleum coke	Nonvolatile solids	Polycyclic structures

Source: Adapted from Morrison and Boyd.[3]

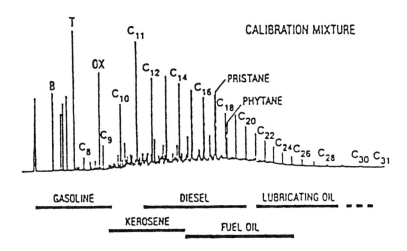

FIGURE 1. Approximate boiling ranges for individual hydrocarbon products. Benzene (B) has a boiling point of 80.1°C and n-Hentri-acontane (C-31) has a boiling point of 302°C. Source: Senn and Johnson (1985).

Physical Properties

Aboveground or in situ remediation of hydrocarbon contaminated soils or waters must address the specific organics present. Cleanup of spilled gasoline is actually the combined cleanup of several individual organic chemical compounds. Each of these organics has specific physical, chemical, and biological properties. First let us summarize the specific compound properties.

Gasoline is a chemically complex petroleum product composed of short branched and straight chain alkanes (paraffins), cycloalkanes, and aromatics.

**Table 2.   The Relative Amount of Paraffins, Cycloparaffins, and Aromatics in the Gasoline
Fraction of Representative Crude Oils**

Origin of Crude Oil	Boiling Range (°C)	Volume (%)		
		Paraffins	Cycloparaffins	Aromatics
Oklahoma (Ponca)	55–180	50	40	10
Pennsylvania	40–200	70	22	8
Texas (Hastings)	50–200	27	67	6
California (Santa Fe Springs)	45–150	41	50	9
Canada (Turner Valley)	45–200	51	35	14
Mexico (Altamira)	40–200	49	36	14
Rumania (Bucsani)	50–150	56	32	12
Kuwait	40–200	72	20	8
Russia (Baku)	60–200	29	63	8

Source: Adapted from Perry.[4]

Table 3. Some of the Major Constituents of the Gasoline Fraction (b.p. 36-117°C) in Selected Petroleums

Constituent	Volume (%)		
	Conroe, TX	Colinga, CA	Jennings, LA
Alkanes:			
n-Pentane	0.33	0.44	1.12
n-Hexane	6.44	7.75	9.15
n-Heptane	6.90	5.94	8.42
2-Methylpentane	2.89	2.56	3.47
2,3-Dimethylhexane	0.22	1.30	2.39
Cycloalkanes:			
Cyclopentane	0.96	1.76	0.67
Methylcyclopentane	6.51	10.29	5.01
Cyclohexane	10.40	7.63	7.13
Methylcyclohexane	22.00	14.55	18.07
Ethylcyclopentane	2.03	4.38	2.34
Trimethylcyclopentane	3.64	8.12	4.18
Aromatics:			
Benzene	3.27	2.22	3.61
Toluene	16.19	7.94	12.02

Source: Adapted from Perry.[4]

Figure 2 provides several examples of these classes of compounds. The molecular weight, density, solubility, boiling point, and vapor pressure for several compounds in gasoline, diesel, and fuel oil are presented in Table 4.

As shown in Table 4, the number of carbon atoms present in a compound has a major effect on its properties. Alkane chains up to 17 carbons in length are liquids and have densities less than that of water (<1). Alkane chains with 18 or more carbons in length are actually solids at room treatment and are commonly referred to as waxes. Alkane solubility rapidly decreases as the number of carbons present in the compound increases. Pentane, with a chain length of five carbons, has a solubility of 360 ppm at 16°C; hexane (with six carbon atoms) has a solubility of 13 ppm and decane (with ten carbon atoms) has a solubility of only 0.009 ppm at 20°C.

Vapor pressures also decrease as alkane carbons numbers increase. High vapor pressures indicate that a compound can easily volatilize; low vapor pressures are associated with chemicals that are semivolatile or nonvolatile. Methane (1 carbon), ethane (2 carbons), propane (3 carbons), and butane (4 carbons) are usually found as gases. For the liquid alkanes previously discussed, pentane has a vapor pressure of 430 mm of Hg at 20°C, hexane of 120 mm of Hg at 20°C, and decane of only 2.7 mm of Hg at 20°C. Boiling point temperatures for alkanes, however, increase with the number of carbons present.

Cycloalkanes are similar to straight or branched chain alkanes in properties. Their densities are less than one, solubilities and vapor pressure decrease with the carbon number, and boiling point temperatures increase with the carbon number.

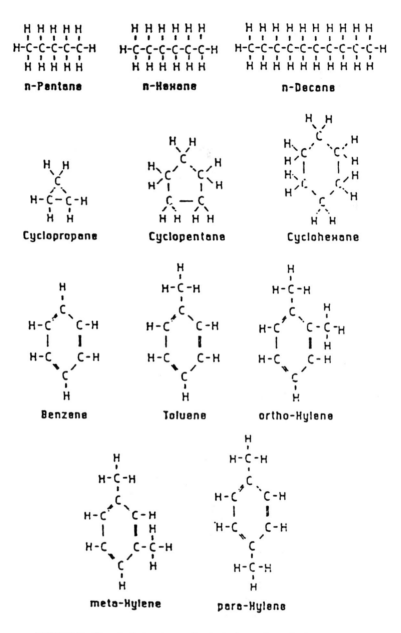

FIGURE 2. Chemical structures of selected petroleum hydrocarbons.

Table 4. Physical/Chemical Properties of Selected Petroleum Hydrocarbons

Compound	Molecular Weight	Density	Solubility (@°C)	Boiling Point, °C	Vapor Pressure @1 atm and (°C)
Pentane	72.15	0.626	360 (16)	36	430 (20)
Hexane	86.17	0.66	13 (20)	68.7	120 (20)
Decane	142.28	0.73	0.009 (20)	173	2.7 (20)
Cyclopropane	42.08	0.72	37,000	-33	760 (-33)
Cyclopentane	70.14	0.751	<1000	—	200 (13.8)
Cyclohexane	84.16	0.779	55 (20)	81	77 (20)
Benzene	78.11	0.878	1780 (20)	80.1	76 (20)
Toluene	92.10	0.867	515 (20)	110.8	22 (20)
ortho-Xylene	106.17	0.88	175 (20)	144.4	5 (20)
meta-Xylene	106.17	0.86	175 (20)	139	6 (20)
para-Xylene	106.17	0.86	198 (25)	138.4	6.5 (20)

Note: Compiled from various sources.

The aromatic fraction of petroleum products is perhaps the most important group of chemicals from an environmental point of view. Benzene, toluene, and the xylenes (BTX) each have densities less than one. Benzene is the most soluble of this class at 1780 ppm at 20°C. Toluene has a solubility of 515 ppm at 20°C. The isomeric xylenes have different solubilities: 175 ppm for ortho and meta-xylene at 20°C and 198 ppm for para-xylene at 25°C. Vapor pressures for these compounds are 76 mm of Hg at 20°C for benzene, 22 mm of Hg at 20°C for toluene, and approximately 6 mm of Hg at 20°C for each of the isomeric xylenes.

REMEDIATION

Now let us look at the effect that different chemical properties have on the remediation of a site. The first thing that we want to know when trying to remediate a site is the location of the contaminants. We know that there are three possible phases in which petroleum hydrocarbons can be found. Hydrocarbons are usually released first to the unsaturated soils. Next, the hydrocarbons travel down through the soils and encounter the aquifer. Hydrocarbons in the vadose zone can also volatilize during movement and contribute to the soil gases. Therefore, we can expect to find spilled petroleum hydrocarbons in the soils, the aquifer, and in the soil gases.

Once petroleum hydrocarbons are introduced into the environment they interact with the surrounding soils. Some of the major processes affecting these chemicals include adsorption, chemical degradation, diffusion, volatilization, and biodegradation. As we have seen, many constituents of petroleum products (such as the alkanes and aromatics) are nonpolar compounds and have only limited solubility in water. Naturally occurring soil compounds, such as humic and fulvic acids, may dissolve in water and help to dissolve other nonpolar compounds. Covalent bonding of contaminants to the functional groups of humic molecules can also serve to immobilize contaminants. In addition, clays in the subsurface frequently have positively charged surfaces that can bind polar as well as charged contaminant molecules.

A common way to compare specific compound migration potentials is to use K_{oc} values. K_{oc} is a measurement of the tendency of an organic compound to be adsorbed by the soil. The higher the K_{oc} value for a compound, the lower its motility and the higher its adsorption potential.[6] Table 5 gives K_{oc} values for some of the compounds found in gasoline. A more complete description of contaminant-soil interactions is beyond the scope of this article. Interested readers can consult soil chemistry texts such as the volume by Dragun.[1]

The problem with remediating a gasoline spill is that we do not find all of the chemical constituents of ''gasoline'' in each of the three different

Table 5. Adsorption Coefficients
for Selected Gasoline
Compounds

Chemical	K_{oc} Value
n-Pentane	568
n-Hexane	1097
n-Heptane	2361
Benzene	50
Toluene	339
ortho-Xylene	255

Source: Adapted from U.S. EPA
1988.[6]

phases previously described. The compounds that make up gasoline and have low solubility, low volatility, and strong adsorption characteristics will be most prevalent with the site soils. The compounds with high solubility will be most prevalent in the aquifer, and the compounds with high volatility will be most prevalent in the soil gases. We tend to have a natural separation of the chemical components of any petroleum product.

Specific chemical properties affect the technologies that are used for remediation as well as the methods used for analysis. We will first review their effects on site remediation.

Soils Remediation

There are four major ways to remediate soils contaminated with petroleum hydrocarbons:

- Excavation and off-site disposal
- In situ soil venting
- In situ biodegradation
- Aboveground or in situ chemical oxidation

Excavation of site soils may result in the loss of the volatile compounds present. As dirt is exposed to the atmosphere, petroleum products with high vapor pressures and low boiling points will tend to volatilize. Some care will have to be taken in areas where air emissions are critical in order not to release large amounts of these compounds. As shown in Table 4, benzene (the most significant compound in terms of human health effects) is expected to be one of the most volatilized compounds.

Next, let us consider in situ soil venting. The components of gasoline will have a major effect on the suitability of this technology. The basis of soil venting is to move air past the contaminated soils and to transfer the organics from a liquid phase into the vapor phase. This mass transfer process effectively removes them from the soil. The rate that the hydrocarbons will vaporize is related to the vapor pressure and the boiling points of the specific compounds. Using the property values listed in Table 4, not all of the gasoline constituents are expected to be amenable to this treatment. Even fewer components of diesel and fuel oil are amenable to soil venting technologies.

Soils can be cleaned by biological methods. All of the compounds found in gasoline, diesel, and fuel oil are degradable by bacteria. However, enhanced bioremediation frequently requires improvements in the subsurface growth environment surrounding the indigenous microorganisms. The site hydrogeologist must help to ensure the transfer and mixing of oxygen and nutrients throughout the contaminated zone. The hydrogeologist must also take care to ensure that soluble and/or volatile components are not further spread through the action of any liquids introduced at the site.

Lastly, chemical oxidation can also be used to remediate hydrocarbon-contaminated soils. This commonly relies on the use of hydrogen peroxide and catalysts to destroy the hydrocarbons present. The treatment process may have to be repeated until all of the regulated hydrocarbon constituents reach acceptable concentrations.

Aquifer Remediation

There are two major strategies employed to remove petroleum hydrocarbons from contaminated aquifers: pump-and-treat in aboveground systems and in situ bioremediation. The properties of the specific organics present will have a significant effect on pump-and-treat methods. Free product floating on the aquifer can be removed through induced groundwater level depression and mechanical pumping. Compounds that are soluble in water can be removed from the subsurface and treated above ground with a variety of technologies. Extremely hydrophobic chemicals may remain adsorbed to subsurface soil particles and escape remediation using pump-and-treat methods.

In situ biodegradation is frequently an effective remediation strategy because all of the compounds are biodegradable. Again, care must be taken to ensure that soluble contaminants are not further mobilized by the subsurface introduction of any liquids.

ANALYZING FOR PETROLEUM HYDROCARBONS

Finally, the variable composition of petroleum products can have a major effect on their identification and quantitation. Chemical analysis for petroleum hydrocarbons is used throughout the site remediation process for several purposes, including:

- Initially, to identify and quantitate the chemicals present
- During the project, to monitor the progress of the remediation effort
- Lastly, to certify that the cleanup effort is complete

Because petroleum products are complex mixtures of chemicals, special problems are faced in analyzing for these compounds. Data interpretation also becomes much more complex. We will point out some of the difficulties of analyzing water and soils for hydrocarbons, as well as some common pitfalls in interpreting this data.

In general, water samples are easier to analyze than soil samples. This is because water samples are more homogeneous than soil samples (assuming no free product is present). Soils tend to be much more heterogeneous in nature, and their actual inorganic and organic composition may change widely over small horizontal or vertical distances. Homogeneity is desired from an analytical standpoint because the sample may be more representative of the site conditions at a given point and time. Heterogeneous samples are more likely to contain organic "hot spots", regions of contamination much higher than typically found.

There are two major types of analyses that can be performed with water or soil samples containing petroleum products. The first type of method is nonspecific and attempts to quantitate the total mass of hydrocarbons present. This method is usually a variation of the "oil and grease" analysis originally performed with waste water. With this method, a given volume of water or wet weight soil sample is extracted with a solvent such as fluorocarbon-113. The total mass of hydrocarbons dissolved in the solvent is then quantitated by comparing the infrared absorption of the extraction liquid against that of a defined hydrocarbon mixture.

The infrared spectrophotometric method previously described, EPA Method 418.1: Petroleum Hydrocarbons, Total Recoverable,[8] has two major drawbacks when used to analyze complex petroleum products. These drawbacks are:

- Volatile compounds are usually lost in the extraction procedure. This applies particularly to the analysis of gasoline contaminated materials.
- Samples are quantitated against a known hydrocarbon mixture (for instance, a mixture of isooctane, n-hexadecane, and chlorobenzene), not the specific petroleum product spilled at the site. All hydrocarbons do not respond equally to infrared analysis, and comparison of the unknown to the standard mixture may result in artificially high or low hydrocarbon concentrations.

Anyone interpreting results from such a test must also consider that:

- All materials (contaminants or benign materials) that are soluble in the solvent will be extracted. These materials may create positive or negative interferences with the hydrocarbon quantitation.

• Soil samples must also be analyzed for moisture content in order to correct the hydrocarbon concentration for the weight of the water present. Remember that a certain wet weight of soil is initially used for extraction. Failure to correct for the percentage of moisture content of the soils results in artificially low hydrocarbon concentration values.

• All extraction and quantitation procedures have a degree of variability. For this method, variability may be 25% or greater depending upon the specific hydrocarbons present and the soil matrix involved. Specific soil spikes may be used to gain a better understanding of the site-specific variability of this technique. The laboratory QA/QC data should be carefully evaluated in interpreting final hydrocarbon concentration values.

The extraction/IR quantitation method does have certain advantages. It is relatively quick and easy to perform and is not very expensive (typically $50 to $75). This technique is perhaps most valuable for use as a screening step in determining the presence of hydrocarbon contamination. It cannot identify and quantitate the specific compounds present, and we do not recommend its use in setting cleanup standards or for certifying that a remediation is complete.

The second method for analyzing hydrocarbons in water and soils involves the specific quantitation of organics with gas chromatography/mass spectrometry (GC/MS). EPA Method SW846-8240 describes a GC/MS purge-and-trap technique for quantitating volatiles present in a sample. Volatile compounds present in a liquid sample are purged using an inert gas and captured on an adsorbent trap. The captured organics are then eluted from the trap into the GC for analysis. Soil samples are rapidly heated and the volatile compounds likewise captured on a trap for further analysis.[7]

EPA Method SW846-8270 describes a GC/MS methylene chloride extraction method for quantitating semivolatiles present in a sample. Soil and liquid samples are extracted in an analogous manner. Unfortunately, a single method is not readily available for identifying and quantitating both volatiles and semivolatiles present in a water or soil sample.[7] To completely characterize a water or soil sample, both the volatile and semivolatile GC/MS analyses should be completed.

GC/MS methods offer the ability to both identify and quantitate the specific organics present. This may be critical if the remediation effort is driven by the desire to reduce only certain compounds, such as benzene, toluene, or xylene. The benefits of GC/MS work have to be weighed against their higher costs. Prices of $100 to $300 per sample are not unusual for either of the volatile or semivolatile GC/MS tests described.

Gas chromatography can also help determine whether a fuel oil sample has been biologically weathered. This is done by comparing the ratio of certain straight chain alkanes ($n\text{-}C_{17}$ and $n\text{-}C_{18}$) with specific branched chain alkanes (pristane and phytane, respectively). The branched chain alkanes are more resistant to biodegradation, and biologically weathered samples typically have low straight chain to branched chain ratios. A more complete description and application of the "pristane/phytane" test is found in the paper by Senn and Johnson.[5]

In general, great care must be taken in deciding which analytical methods are used for identifying and quantitating petroleum hydrocarbons present in soils and liquids. Equally important is the careful evaluation of data generated by these tests. Just because a number is presented in an official-looking report does not mean that it is correct or even representative of the overall conditions at your site.

SUMMARY

As the reader can see, the properties of specific organic chemicals present in a complex petroleum product have a major effect on the distribution of the compounds in a soil/gas/liquid matrix. The initial mixture of the compounds will tend to separate and equilibriate in the soil and aquifer environments. The volatile compounds will tend to be found in the soil gases and the atmosphere. The nonsoluble compounds will tend to stay with the soil particles, and the soluble compounds will tend to dissolve and travel with the water moving through the contaminated area. However, many complex reactions can take place between contaminants and the subsurface environment to affect these generalizations.

The identification and quantitation of petroleum products in soils and ground waters is a complex and difficult task. General procedures measuring the total mass of extractable hydrocarbons can provide useful site information, but the limitations of these tests must be considered when evaluating the resultant analytical data. Gas chromatography with mass spectroscopy provides a more definitive method of identifying and quantitating hydrocarbons present in soil and liquid samples. However, the advantages of using these methods are frequently (and, in many cases, shortsightedly) overcome by their higher costs.

Lastly, we would like to make one other observation based upon the data provided in this chapter. Because the individual hydrocarbon contaminants partition unevenly between site soils, waters, and gases, a single type of remediation technology may not be able to efficiently and cost-effectively remove all of the components present at a site. Currently available remediation technologies (such as soil excavation, soil venting, pump-and-treat, in situ bioremediation, etc.) tend to work best only against certain specific compounds. There is no single remedial method that is best for all classes of compounds under all types of site conditions. Successful remediation efforts may have to rely on the proper application of a combination of remediation technologies.

REFERENCES

1. **Dragun, J.** *The Soil Chemistry of Hazardous Materials,* (Silver Springs, MD: Hazardous Materials Control Research Institute, 1988).
2. **Hawley, G.** *The Condensed Chemical Dictionary,* 10th ed. (New York: Van Nostrand Reinhold Co., Inc., 1981).
3. **Morrison, R. T. and R. N. Boyd.** *Organic Chemistry,* 3rd ed. (Boston, MA: Allyn and Bacon, Inc., 1973).
4. **Perry, J. J.** "Microbial metabolism of cyclic alkanes," in *Petroleum Microbiology,* R. M. Atlas, Ed. (New York: Macmillan Publishing Co., 1984, pp. 61-98).
5. **Senn, R. B. and M. S. Johnson.** "Interpretation of gas chromatography data as a tool in subsurface hydrocarbon investigations," in *Proceedings of the NWWA/API Conference on Petroleum Hydrocarbons and Organic Chemicals in Ground Water — Prevention, Detection and Restoration.* Houston, Texas, November 13-15, 1985. National Water Well Association, Dublin, Ohio.
6. "Cleanup of Releases from Petroleum USTs: Selected Technologies." EPA/530/UST-88/001. United States Environmental Protection Agency. (1988).
7. "Test Methods for Evaluating Solid Waste," vol. 1B, 3rd ed. SW846. United States Environmental Protection Agency. (1986).
8. "Methods for Chemical Analysis of Water and Wastes." EPA-600/4-79-020. United States Environmental Protection Agency. (1983).

CHAPTER 3

Using the Properties of Organic Compounds to Help Design a Treatment System

Evan Nyer, Gary Boettcher, and Bridget Morello

This chapter will provide the physical/chemical and treatability properties of 50 compounds. These physical/chemical parameters can be used to help evaluate data generated during remedial investigations. The treatability parameters can be used as a basis for the preliminary design of a treatment system that will remove organic compounds from ground water.

The biggest obstacle in designing a treatment system is where to begin. Typically, the two main starting points I have seen applied are laboratory treatability studies and "by the book" design. Neither of these methods is accurate or efficient. In laboratory treatability studies the designer generally submits a groundwater sample to the laboratory for purposes of simulating full scale treatment units. Laboratory treatability studies, however, cannot be used as a direct simulation of most organic treatment processes. Textbooks should never be used as "cookbooks" for the design of a treatment system.

ISBN 0-87371-731-7

The cookbook recipe simply uses every treatment method available for removing organic compounds and sizes unit operations based on values supplied in the textbook. The final design utilizes all the treatment units in series. Textbooks, including my own, should be used for general knowledge and reference purposes only, not for design data.

The treatment system designs I have worked on have always been preceded by complete evaluation of the properties of the compounds. While I would not proceed directly to a full-scale installation based strictly upon analysis of compound properties, they can provide several insights for final design. Most importantly, the properties of compounds can indicate critical points of a design and areas requiring further data. These areas can then be further evaluated in laboratory and field pilot tests.

The main physical/chemical properties that should be evaluated prior to design are solubility, specific gravity, and octanol/water coefficient. These properties mainly help us understand data generated during remedial investigations. But, they will have some input in the following treatment system.

SOLUBILITY

Solubility is one of the most important properties affecting the fate and transport of organic compounds in the environment. The solubility of a compound is described as the maximum dissolved quantity of compound in pure water at a specific temperature. Solubilities of most common organic compounds range from 1 to 100,000 ppm at ambient treatment. However, several compounds exhibit higher solubilities and some are infinitely soluble. Highly soluble compounds are easily transported by the hydrologic cycle. They tend to have low adsorption coefficients for soils and low bioconcentration factors in aquatic life. Highly soluble compounds also tend to be more readily biodegradable.

Solubility usually decreases as temperature increases due to an increase in water vapor pressure at the liquid/gas interface. Escaping molecules then force larger numbers of gas molecules out of solution. Table 1 presents the solubility values for 50 organic compounds.

When reviewing the results from a groundwater study the concentrations of organic compounds should be related to the solubilities of those compounds. For example, high concentrations of nonsoluble compound may indicate the presence of a pure compound DNAPL. Therefore, the treatment system should be designed with the capability to treat pure compounds.

SPECIFIC GRAVITY

Specific gravity is a dimensionless parameter derived from density. The specific gravity of a compound is defined as the ratio of the weight of the compound of a given volume and at a specified temperature to the weight of the same volume of water at a given temperature. The specific gravity of water at 4°C is usually used as a basis because the density of water at 4°C is 1.000 g/mL.

In environmental analysis the primary reason for knowing the specific gravity of a compound is to determine whether liquids will float or sink in water. Table 2 presents the specific gravities of 50 organic compounds. Pure compounds that are lighter than water will form a layer on top of the water. Organic compounds that are heavier than water will move through the aquifer until they are fully adsorbed by soil properties or until they encounter an impenetrable layer.

OCTANOL/WATER PARTITION COEFFICIENT

The octanol/water partition coefficient (K_{ow}) is defined as the ratio of a compound's concentration in the octanol phase to its concentration in the aqueous phase of a two-phase system. Measured values for organic compounds range from 10^{-3} to 10^7. Low K_{ow} values (<10) are considered hydrophilic and tend to have higher water solubility. High K_{ow} values ($>10^4$) are very hydrophobic.

K_{ow} values for organic compounds are used to evaluate fate in the environment. The parameter can be related to solubility in water and bioconcentration effects, but it is mainly used to relate to soil/sediment adsorption. Table 3 presents K_{ow} values for 50 organic compounds. When combined with the organic content of the soil, these values can be used to predict the amount of material adsorbed in the soil and the retardation factor for movement through the aquifer.

When pure compounds are lost to the environment it is important to know where they are likely to be found. Soluble compounds will migrate with the surface water which will infiltrate the aquifer and migrate with the ground water. Nonsoluble compounds will be adsorbed on the soil. However, if the mass of organic compounds exceeds the adsorptive capacity of the soil, the compounds will continue to migrate until they reach the aquifer. Compounds with low specific gravity will be retained at the surface of the aquifer; compounds with high specific gravity will continue to move vertically through the aquifer.

The physical/chemical properties presented here will help the reader understand where compounds of concern should be in the ground and/or aquifer.

Table 1. Solubility for Specific Organic Compounds

	Compound	Solubility[a] (mg/L)	Ref.
1	Acenaphthene	3.42	2
2	Acetone	1×10^{6a}	1
3	Aroclor 1254	1.2×10^{-2}	2
4	Benzene	1.75×10^3	1a
5	Benzo(a)pyrene	1.2×10^{-3}	2
6	Benzo(g,h,i)perylene	7×10^{-4}	2
7	Benzoic Acid	2.7×10^3	2
8	Bromodichloromethane	4.4×10^3	2
9	Bromoform	3.01×10^3	1b
10	Carbon Tetrachloride	7.57×10^2	1a
11	Chlorobenzene	4.66×10^2	1a
12	Chloroethane	5.74×10^3	2
13	Chloroform	8.2×10^3	1a
14	2-Chlorophenol	2.9×10^4	2
15	p-Dichlorobenzene (1,4)	7.9×10^1	2
16	1,1-Dichloroethane	5.5×10^3	1a
17	1,2-Dichloroethane	8.52×10^3	1a
18	1,1-Dichloroethylene	2.25×10^3	1a
19	cis-1,2-Dichloroethylene	3.5×10^3	1a
20	trans-1,2-Dichloroethylene	6.3×10^3	1a
21	2,4-Dichlorophenoxyacetic Acid	6.2×10^2	2
22	Dimethyl Phthalate	4.32×10^3	2
23	2,6-Dinitrotoluene	1.32×10^3	2
24	1,4-Dioxane	4.31×10^5	2
25	Ethylbenzene	1.52×10^2	1a
26	bis(2-Ethylhexyl)phthalate	2.85×10^{-1}	2
27	Heptachlor	1.8×10^{-1}	2
28	Hexachlorobenzene	6×10^{-3}	1a
29	Hexachloroethane	5×10^1	2
30	2-Hexanone	1.4×10^4	2
31	Isophorone	1.2×10^4	2
32	Methylene Chloride	2×10^4	1
33	Methyl Ethyl Ketone	2.68×10^5	1b
34	Methyl Naphthalene	2.54×10^1	2a
35	Methyl tert-Butyl Ether	4.8	3
36	Naphthalene	3.2×10^1	2
37	Nitrobenzene	1.9×10^3	2
38	Pentachlorophenol	1.4×10^1	1
39	Phenol	9.3×10^4	1a,b
40	1,1,2,2-Tetrachloroethane	2.9×10^3	2
41	Tetrachloroethylene	1.5×10^2	1a
42	Tetrahydrofuran	3×10^{-1}	4
43	Toluene	5.35×10^2	1a
44	1,2,4-Trichlorobenzene	3×10^1	2
45	1,1,1-Trichloroethane	1.5×10^3	1a
46	1,1,2-Trichloroethane	4.5×10^3	1a
47	Trichloroethylene	1.1×10^3	1a
48	2,4,6-Trichlorophenol	8×10^2	2
49	Vinyl Chloride	2.67×10^3	1a
50	o-Xylene	1.75×10^2	1c

[a] Solubility of 1,000,000 mg/L assigned because of reported "infinite solubility" in the literature.

1. Superfund Public Health Evaluation Manual, Office of Emergency and Remedial Response Office of Solid Waste and Emergency Response, U.S. Environmental Protection Agency. (1986).
   a. Environmental Criteria and Assessment Office (ECAO), EPA, Health Effects Assessments for Specific Chemicals. (1985).
   b. **Mabey, W. R., Smith, J. H., Rodoll, R. T., Johnson, H. L., Mill, T., Chou, T. W., Gates, J., Patridge, I. W., Jaber, H., and Vanderberg, D.,** "Aquatic Fate Process Data for Organic Priority Pollutants", EPA Contract Nos. 68-01-3867 and 68-03-2981 by SRI International, for Monitoring and Data Support Division, Office of Water Regulations and Standards, Washington, D.C. (1982).
   c. **Dawson, et al.,** "Physical/Chemical Properties of Hazardous Waste Constituents," by Southeast Environmental Research Laboratory for USEPA. (1980).
2. USEPA "Basics of Pump-and-Treat Groundwater Remediation Technology" EPA/600/8-901003, Robert S. Kerr Environmental Research Laboratory. (March 1990).
3. Manufacturer's data; Texas Petrochemicals Corporation, Gasoline Grade Methyl tert-butyl ether Shipping Specification and Technical Data. (1986).
4. *CRC Handbook of Chemistry and Physics,* 71st ed. (Boca Raton, FL: CRC Press, Inc., 1990).

These properties are also necessary for use in designing treatment systems such as oil/water separators and liquid/liquid extractors.

The main treatability parameters that should be used to help design a treatment system are strippability (Henry's Law constant), adsorbability, and biodegradability. These parameters are discussed in the following paragraphs.

HENRY'S LAW CONSTANT

Generally, for non-ideal solutions, Henry's Law states that the equilibrium partial pressure of a compound in the air above the air/water interface is proportional to the concentration of that compound in the water. Henry's Law can be expressed as follows:

$$P_A = H_A x_A$$

where: P_A = Partial pressure of a compound in liquid at equilibrium with gas (atm)
H_A = Henry's Law constant (atm)
x_A = Mole fraction of a compound in gas (mole/mole)

Therefore, Henry's Law constant expresses the amount of chemical partitioning between air and water at equilibrium.

Table 2. Specific Gravity for Specific Organic Compounds

	Compound	Specific[a] Gravity	Ref.
1	Acenaphthene	1.069 (95°/95°)	1
2	Acetone	.791	1
3	Aroclor 1254	1.5 (25°)	3
4	Benzene	.879	1
5	Benzo(a)pyrene	1.35 (25°)	4
6	Benzo(g,h,i)perylene	NA	
7	Benzoic Acid	1.316 (28°/4°)	1
8	Bromodichloromethane	2.006 (15°/4°)	1
9	Bromoform	2.903 (15°)	1
10	Carbon Tetrachloride	1.594	1
11	Chlorobenzene	1.106	1
12	Chloroethane	.903 (10°)	1
13	Chloroform	1.49 (20°C liquid)	2
14	2-Chlorophenol	1.241 (18.2°/15°)	1
15	p-Dichlorobenzene (1,4)	1.458 (21°)	1
16	1,1-Dichloroethane	1.176	1
17	1,2-Dichloroethane	1.253	1
18	1,1-Dichloroethylene	1.250 (15°)	1
19	cis-1,2-Dichloroethylene	1.27 (25°C liquid)	2
20	trans-1,2-Dichloroethylene	1.27 (25°C liquid)	2
21	2,4-Dichlorophenoxyacetic Acid	1.255	6
22	Dimethyl Phthalate	1.189 (25°/25°)	1
23	2,6-Dinitrotoluene	1.283 (111°)	1
24	1,4-Dioxane	1.034	1
25	Ethylbenzene	.867	1
26	bis(2-Ethylhexyl)phthalate	.9843	1
27	Heptachlor	1.57	5
28	Hexachlorobenzene	2.044	1
29	Hexachloroethane	2.09	6
30	2-Hexanone	.815 (18°/4°)	1
31	Isophorone	.921 (25°)	2
32	Methylene Chloride	1.366	1
33	Methyl Ethyl Ketone	.805	1
34	Methyl Naphthalene	1.025 (14°/4°)	1
35	Methyl tert-Butyl Ether	.731	1
36	Naphthalene	1.145	1
37	Nitrobenzene	1.203	1
38	Pentachlorophenol	1.978 (22°)	1
39	Phenol	1.071 (25°/4°)	1
40	1,1,2,2-Tetrachloroethane	1.600	1
41	Tetrachloroethylene	1.631 (15°/4°)	1
42	Tetrahydrofuran	.888 (21°/4°)	1
43	Toluene	.866	1
44	1,2,4-Trichlorobenzene	1.446 (26°)	1
45	1,1,1-Trichloroethane	1.346 (15°/4°)	1
46	1,1,2-Trichloroethane	1.441 (25.5°/4°)	1
47	Trichloroethylene	1.466 (20°/20°)	1
48	2,4,6-Trichlorophenol	1.490 (75°/4°)	1
49	Vinyl Chloride	.908 (25°/25°)	1
50	o-Xylene	.880	1

[a] Specific gravity of compound at 20°C referred to water at 4°C (20°/4°) unless otherwise specified.

NA = Not Available

1. **Dean, J. A.** *Lange's Handbook of Chemistry,* 11th ed. (New York: McGraw-Hill Book Co., 1973).
2. **Weiss, G.** *Hazardous Chemicals Data Book,* 2nd ed. (New York: Noyes Data Corp., 1986).
3. "Draft Toxicological Profile for Selected PCBs" (November 1987). *U.S. Public Health Service Agency for Toxic Substances and Disease Registry.*
4. "Draft Toxicological Profile for Benzo(a)pyrene" (October 1987). *U.S. Public Health Service Agency for Toxic Substances and Disease Registry.*
5. **Verschueren, K.** *Handbook of Environmental Data on Organic Chemicals,* 2nd ed. (New York: Van Nostrand Reinhold Co., 1983).
6. Merck Index, 9th ed. (Rahway, NJ: Merck and Co., Inc., 1976).

Aeration is a technology often employed in water treatment applications to strip the concentration of volatile organic compounds (VOCs) from water. The controlling factor for removal of VOCs from water is the rate of transfer from the liquid phase (water) to the gas phase (air) until equilibrium is established. The transfer rate of VOCs from water via aeration depends upon the treatment of both the water and the air, as well as the physical and chemical properties of the VOCs. Temperature changes of as little as 10°C can result in threefold increases in Henry's Law constants. In a gas-liquid system, the equilibrium vapor concentration of a VOC can be computed from the compound specific Henry's Law constant and total system pressure.

Generally, the greater the Henry's Law constant (i.e., greater than 160 atm), the more volatile a compound and the more easily it can be removed from solution. Henry's Law constants can be plugged into a computer model to develop a preliminary design and cost estimate for an air stripper. Table 4 presents Henry's Law constants for 50 organic compounds.

CARBON ADSORPTION CAPACITY

Activated carbon has variable effectiveness adsorbing organic compounds. Low molecular weight, polar compounds are not well adsorbed. High molecular weight, nonpolar compounds, such as pesticides, polychlorinated biphenyls, phthalates, and aromatics, are readily adsorbed.

Activated carbon adsorption isotherm data can be used to evaluate the carbon adsorptive capacity for organic compounds. These data may be used to complete an initial estimate of the organic mass that carbon will adsorb. Since the main cost of carbon adsorption is carbon, this mass data can be used as a preliminary basis for cost estimation. Table 5 presents carbon adsorption capacity values for 50 organic compounds.

Table 3. Octanol Water Coefficients (K_{ow}) for Specific Organic Compounds

	Compound	K_{ow}	Ref.
1	Acenaphthene	1.0×10^4	2
2	Acetone	6×10^{-1}	1d
3	Aroclor 1254	1.07×10^6	2
4	Benzene	1.3×10^2	1a
5	Benzo(a)pyrene	1.15×10^6	2
6	Benzo(g,h,i)perylene	3.24×10^6	2
7	Benzoic Acid	7.4×10^1	2
8	Bromodichloromethane	7.6×10^1	2
9	Bromoform	2.5×10^2	1b
10	Carbon Tetrachloride	4.4×10^2	1a
11	Chlorobenzene	6.9×10^2	1a
12	Chloroethane	3.5×10^1	2
13	Chloroform	9.3×10^1	1a
14	2-Chlorophenol	1.5×10^1	2
15	p-Dichlorobenzene (1,4)	3.9×10^3	2
16	1,1-Dichloroethane	6.2×10^1	1a
17	1,2-Dichloroethane	3.0×10^1	1a
18	1,1-Dichloroethylene	6.9×10^1	1a
19	cis-1,2-Dichloroethylene	5.0	1a
20	trans-1,2-Dichloroethylene	3.0	1a
21	2,4-Dichlorophenoxyacetic Acid	6.5×10^2	2
22	Dimethyl Phthalate	1.3×10^2	2
23	2,6-Dinitrotoluene	1.0×10^2	2
24	1,4-Dioxane	1.02	2
25	Ethylbenzene	1.4×10^3	1a
26	bis(2-Ethylhexyl)phthalate	9.5×10^3	2
27	Heptachlor	2.51×10^4	2
28	Hexachlorobenzene	1.7×10^5	1a
29	Hexachloroethane	3.98×10^4	2
30	2-Hexanone	2.5×10^1	3
31	Isophorone	5.0×10^1	2
32	Methylene Chloride	1.9×10^1	1b
33	Methyl Ethyl Ketone	1.8	1a
34	Methyl Naphthalene	1.3×10^4	2
35	Methyl tert-Butyl Ether	NA	
36	Naphthalene	2.8×10^3	2
37	Nitrobenzene	7.1×10^1	2
38	Pentachlorophenol	1.0×10^5	1b
39	Phenol	2.9×10^1	1a
40	1,1,2,2-Tetrachloroethane	2.5×10^2	2
41	Tetrachloroethylene	3.9×10^2	1a
42	Tetrahydrofuran	6.6	4
43	Toluene	1.3×10^2	1a
44	1,2,4-Trichlorobenzene	2.0×10^4	2
45	1,1,1-Trichloroethane	3.2×10^2	1b
46	1,1,2-Trichloroethane	2.9×10^2	1a
47	Trichloroethylene	2.4×10^2	1a
48	2,4,6-Trichlorophenol	7.4×10^1	2
49	Vinyl Chloride	2.4×10^1	1a
50	o-Xylene	8.9×10^2	1c

NA = Not Available

1. Superfund Public Health Evaluation Manual, Office of Emergency and Remedial Response Office of Solid Waste and Emergency Response, U.S. Environmental Protection Agency. (1986).
   a. Environmental Criteria and Assessment Office (ECAO), EPA, Health Effects Assessments for Specific Chemicals. (1985).
   b. **Mabey, W. R., Smith, J. H., Rodoll, R. T., Johnson, H. L., Mill, T., Chou, T. W., Gates, J., Patridge, I. W., Jaber, H., and Vanderberg, D.**, "Aquatic Fate Process Data for Organic Priority Pollutants", EPA Contract Nos. 68-01-3867 and 68-03-2981 by SRI International, for Monitoring and Data Support Division, Office of Water Regulations and Standards, Washington, D.C. (1982).
   c. **Dawson, et al.**, "Physical/Chemical Properties of Hazardous Waste Constituents," by Southeast Environmental Research Laboratory for USEPA. (1980).
   d. *Handbook of Environmental Data for Organic Chemicals,* 2nd ed., (New York: Van Nostrand Reinhold Co., 1983).
2. USEPA "Basics of Pump-and-Treat Ground-Water Remediation Technology", EPA/600-8-90/003, Robert S. Kerr Environmental Research Laboratory. (March 1990).
3. **Lyman, W. J., et al.** "Research and Development of Methods for Estimating Physicochemical Properties of Organic Compounds of Environmental Concern". (June 1981).
4. EPA Draft Document "Hazardous Waste Treatment, Storage and Disposal Facilities (TSDF) Air Emissions Model". (April 1989).

BIODEGRADABILITY

Organic compounds are transformed by biochemical reactions in the environment and in engineered unit operations. Biodegradation of organic compounds occurs both aerobically and/or anaerobically depending on the molecular structure of the chemical and the environmental conditions. Engineered bioremediation is necessary to enhance natural processes that are usually less than optimal in the environment.

The first and most important parameters to evaluate before implementing bioremediation is determining whether the compound is degradable, the most effective biodegradation mechanism (aerobic vs anaerobic), and the biodegradation rate. From an ecological point of view, chemicals that are completely degradable, but slow, can be persistent in the environment for a long period of time.

Categorizing biodegradation potential has been reviewed and can be categorized as degradable, persistent, and recalcitrant. Readily degradable refers to compounds that have passed biodegradability tests in a variety of aerobic environments. Degradable also refers to compounds that are normally degraded in tests but not necessarily in the environment. Persistent refers to

Table 4. Henry's Law Constants for Specific Organic Compounds

	Compound	Henry's Law Constant[a] atm	Ref.
1	Acenaphthene	5.1	5
2	Acetone	0	1
3	Aroclor 1254	150	5
4	Benzene	230	1
5	Benzo(a)pyrene	.1	5
6	Benzo(g,h,i)perylene	0	5
7	Benzoic Acid	0	5
8	Bromodichloromethane	127	1
9	Bromoform	35	3
10	Carbon Tetrachloride	1282	1
11	Chlorobenzene	145	2
12	Chloroethane	34	5
13	Chloroform	171	1
14	2-Chlorophenol	0.93	2
15	p-Dichlorobenzene(1,4)	104	4
16	1,1-Dichloroethane	240	1
17	1,2-Dichloroethane	51	1
18	1,1-Dichloroethylene	1841	1
19	cis-1,2-Dichloroethylene	160	1
20	trans-1,2-Dichloroethylene	429	1
21	2,4-Dichlorophenoxyacetic Acid	10	5
22	Dimethyl Phthalate	0	5
23	2,6-Dinitrotoluene	.2	5
24	1,4-Dioxane	.6	5
25	Ethylbenzene	359	1
26	bis(2-Ethylhexyl)phthalate	0	5
27	Heptachlor	46	5
28	Hexachlorobenzene	37.8	2
29	Hexachloroethane	138	5
30	2-Hexanone	1.6	5
31	Isophorone	.3	5
32	Methylene Chloride	89	1
33	Methyl Ethyl Ketone	1.16	2
34	Methyl Naphthalene	3.2	2
35	Methyl tert-Butyl Ether	196	1
36	Naphthalene	20	4
37	Nitrobenzene	1.2	5
38	Pentachlorophenol	0.15	2
39	Phenol	0.017	2
40	1,1,2,2-Tetrachloroethane	21	5
41	Tetrachloroethylene	1035	1
42	Tetrahydrofuran	2	5
43	Toluene	217	1
44	1,2,4-Trichlorobenzene	128	5
45	1,1,1-Trichloroethane	390	1
46	1,1,2-Trichloroethane	41	2
47	Trichloroethylene	544	1
48	2,4,6-Trichlorophenol	.2	5
49	Vinyl Chloride	355000	3
50	o-Xylene	266	1

[a] at water temperature of 68°F

1. Hydro Group, Inc., 1990.
2. **Verschueren, K.**, Solubility and vapor phase pressure data, *Handbook of Environmental Data on Organic Chemicals,* 2nd ed., (New York: Van Nostrand Reinhold Co., 1983).
3. **Kavanaugh, M. C. and Trussel, R. R.**, "Design of Aeration Towers to Strip Volatile Contaminants from Drinking Water", *J. Am. Water Works Assoc.* (December 1980), p. 685.
4. **Yurteri, C., Ryan, D. F., Callow, J. J., and Gurol, M. D.**, "The Effect of Chemical Composition of Water on Henry's Law Constant", *J. Water Pollut. Control Fed.* 59 (11): 954, (November 1987).
5. USEPA "Basics of Pump-and-Treat Ground-water Remediation Technology", EPA/600-8-90/003, Robert S. Kerr Environmental Research Laboratory. (March 1990).

chemicals that remain in the environment for long periods of time. These compounds are not necessarily "nondegradable", but degradation requires long periods of acclimation or modification of the environment to induce degradation.

From the literature, each compound must be evaluated to determine the estimated time to complete the transformation of the chemical under optimal conditions. If the time period is acceptable, treatability and pilot plants can then be initiated. Table 6 presents biodegradation potential for 50 organic compounds.

We can combine these treatability properties with our experience in full-scale design and generate a theoretical preliminary design. This design can be used to generate a preliminary cost estimate. Based upon this data, we can eliminate the technologies that obviously will not work. This data will also show us which compounds are controlling the designs. We can then go back and confirm their concentrations in the field, and test the actual treatment in laboratory and pilot plant tests.

I hope the reader finds these tables to be a convenient source of important information. Please do not use the data as a final basis for full-scale design.

Table 5. Adsorption Capacity for Specific Organic Compounds

	Compound	Adsorption Capacity (mg compound/g carbon) at 500 ppb	Ref.
1	Acenaphthene	155	4
2	Acetone	43	1
3	Aroclor 1254	NA	
4	Benzene	80	1
5	Benzo(a)pyrene	24.8	4
6	Benzo(g,h,i)perylene	8.3	4
7	Benzoic Acid	40 (at pH = 3)	4
8	Bromodichloromethane	5	4
9	Bromoform	13.6	4
10	Carbon Tetrachloride	6.2	2
11	Chlorobenzene	45	3
12	Chloroethane	0.3	4
13	Chloroform	1.6	1
14	2-Chlorophenol	38	3
15	p-Dichlorobenzene (1,4)	87.3	4
16	1,1-Dichloroethane	1.2	4
17	1,2-Dichloroethane	2	2
18	1,1-Dichloroethylene	3.4	4
19	cis-1,2-Dichloroethylene	9	5
20	trans-1,2-Dichloroethylene	2.2	4
21	2,4-Dichlorophenoxyacetic Acid	NA	
22	Dimethyl phthalate	91.2	4
23	2,6-Dinitrotoluene	116	4
24	1,4-Dioxane	0.5 − 1.0	5
25	Ethylbenzene	18	1
26	bis(2-Ethylhexyl)phthalate	3995	4
27	Heptachlor	631.5	4
28	Hexachlorobenzene	42	3
29	Hexachloroethane	74.2	4
30	2-Hexanone	<13	5
31	Isophorone	24.4	4
32	Methylene Chloride	0.8	3
33	Methyl Ethyl Ketone	94	1
34	Methyl Naphthalene	150	5
35	Methyl tert-Butyl Ether	6.5	5
36	Naphthalene	5.6	3
37	Nitrobenzene	50.5	4
38	Pentachlorophenol	100	3
39	Phenol	161	1
40	1,1,2,2-Tetrachloroethane	8.2	4
41	Tetrachloroethylene	34.5	2
42	Tetrahydrofuran	<0.5	5
43	Toluene	50	1
44	1,2,4-Trichlorobenzene	126.6	4
45	1,1,1-Trichloroethane	2	2
46	1,1,2-Trichloroethane	3.7	4
47	Trichloroethylene	18.2	2
48	2,4,6-Trichlorophenol	179 (at pH = 3)	4
49	Vinyl Chloride	TRACE	3
50	o-Xylene	75	4

NA = Not Available

1. **Verschuren, K.** "Handbook of Environmental Data on Organic Chemicals" (New York: Van Nostrand Reinhold, 1983).
2. **Uhler, R. E., et al.** "Treatment Alternatives for Groundwater Contamination." James M. Montgomery, Consulting Engineers.
3. **Stenzel, M.** Letter of Correspondence to Evan Nyer. (August 22, 1984).
4. **USEPA** "Carbon Adsorption Isotherms for Toxic Organics", EPA-600/8-80-023, Municipal Environmental Research Laboratory. (April 1980).
5. **Roy, A.** Calgon Carbon. (1991).

Table 6. Disappearance or Biodegradation Potential for Specific Organic Compounds

	Compound	Biodegradability	Ref.
1	Acenaphthene	D	2
2	Acetone	D	
3	Aroclor 1254	P,D	2,3
4	Benzene	D	1
5	Benzo(a)pyrene	P,D	2,4
6	Benzo(g,h,i)perylene	P,D	2,4
7	Benzoic Acid	D	2
8	Bromodicloromethane	P,D	1
9	Bromoform	P,D	1
10	Carbon Tetrachloride	P,D	1
11	Chlorobenzene	D	1
12	Chloroethane	D	6
13	Chloroform	P,D	1
14	2-Chlorophenol	D	1
15	p-Dichlorobenzene (1,4)	P,D	1
16	1,1-Dichloroethane	P,D	1
17	1,2-Dichloroethane	P,D	1
18	1,1-Dichloroethylene	P,D	1
19	cis-1,2-Dichloroethylene	P,D	1
20	trans-1,2-Dichloroethylene	P,D	1
21	2,4-Dichlorophenoxyacetic Acid	D	2
22	Dimethyl Phthalate	D	5
23	2,6-Dinitrotoluene	D,P	2,5
24	1,4-Dioxane	P,R	8
25	Ethylbenzene	D	1
26	bis(2-ethylhexyl)phthalate	P,D	2,5
27	Heptachlor	P,R	2
28	Hexachlorobenzene	P,R	1
29	Hexachloroethane	D	2
30	2-Hexanone	D	5
31	Isophorone	D	5
32	Methylene Chloride	D	1
33	Methyl Ethyl Ketone	D	
34	Methyl Naphthalene	D	1
35	Methyl tert-Butyl Ether	NA	
36	Naphthalene	D	1
37	Nitrobenzene	D	2
38	Pentachlorophenol	P,D	1
39	Phenol	D	1
40	1,1,2,2-Tetrachloroethane	P,D	2,5
41	Tetrachloroethylene	P,D	1
42	Tetrahydrofuran	D	7
43	Toluene	D	1
44	1,2,4-Trichlorobenzene	P,D	2
45	1,1,1-Trichloroethane	P,D	1
46	1,1,2-Trichloroethane	P,D	1
47	Trichloroethylene	P,D	1
48	2,4,6-Trichlorophenol	D	2
49	Vinyl Chloride	P,D	1
50	o-Xylene	D	1

D	= Degradable	R	= Recalcitrant
P	= Persistant	NA	= Not Available

1. Citations compiled in Nyer, E. K. *Groundwater Treatment Technology,* 2nd ed., in production.
2. **Dragun, J.** *The Soil Chemistry of Hazardous Materials,* The Hazardous Materials Control Research Institute, pp. 367-377 (1988).
3. **Bedard, D. L.** "Bacterial Transformations of Polychlorinated Biphenyls", in *Biotechnology and Biodegradation,* Kanely, D., Chakrabarty, A., and Omenn, G. S., Eds. *Advances in Applied Biotechnology Series, Vol. 4,* (The Woodlands, TX: Portfolio Pub. Co., 1990).
4. "Characterization and Laboratory Soil Treatability Studies for Creosote and Pentachlorophenol Sludges and Contaminated Soil", EPA: Washington, D.C., EPA/600/2-88/055 (1988).
5. **Pitter, P. and Chudoba, J.** *Biodegradability of Organic Substances in the Aquatic Environment,* (Boca Raton, FL: CRC Press, 1990).
6. **Vogel, T. M. and McCarty, P. L.** "Transformations of Halogenated Aliphatic Compounds", *Environ. Sci. Technol.,* 21, 722-736 (1987).
7. **Volskay, V. T. and Grady, C. P.** "Toxicity of Selected RCRA Compounds to Activated Sludge Microorganisms", *J. Water Pollut. Control Fed.* 60 (10): 1850 (1988).
8. **Klecka, G. M. and Gonsoir, S. J.** "Removal of 1,4-Dioxane from Wastewater", *J. Hazard. Mat.,* 13, 161-168 (1986).

CHAPTER 4

The Effects of Biochemical Reactions on Investigations and Remediations

Evan K. Nyer

I would like to try and make everyone more aware of the broad application that bacteria have in many things that we do as professionals. Bacteria and biological methods are having a large impact on the soils and groundwater areas.

Most groundwater professionals tend to think of biological methods as being limited to in situ treatment of organic contamination. We regard bacteria as a state-of-the-art method of remediation. We know bacterial activity is hard to model, and most of us have had little direct contact with biological methods. Biological treatment ends up being thought to be more art than science.

These are the types of reactions that I usually run into when dealing with a biological effect on an investigation or a remediation. Personally, I have

ISBN 0-87371-731-7
© 1992 by Lewis Publishers

worked with bacteria for more than 15 years. I have worked on everything from development of specialized bacteria to aboveground treatment equipment and in situ cleanups. I have worked on biological treatment of soil, water, and air streams. These projects have provided me with a working knowledge and, more importantly, a comfort level with biological systems.

Biological methods are simple and straightforward. Let me share some of my thinking in specific areas of investigation and remediation. Hopefully, as we break these down into their components, we will be able to see their simplicity.

INVESTIGATION

There are four areas that I would like to cover under investigation. We usually think that the only time that bacteria matters in an investigation is when we are evaluating a possible in situ cleanup. However, the bacteria naturally occur in the soil and their activity will affect our investigation. There are four obvious areas in which bacteria can affect our efforts. They are transformation, soil chemistry/biochemistry, modeling, and risk assessment.

Transformation

Let me start with an area that is familiar to many people. Natural bacteria in the soil can transform the chemical nature of a compound. The organic that you find in the ground water may not be the same compound that you spilled on the ground. Bacteria do not always ingest an organic compound and produce CO_2 and new bacteria. The enzymes that they produce cause one reaction at a time. The original contaminant may only encounter one of these enzymes.

Most professionals encounter transformation associated with chlorinated hydrocarbons. Most of the vinyl chloride that is found in the ground and ground water has been transformed from other compounds. Trichloroethylene (TCE) and the various forms of trichloroethane can lose one or two of their chlorine molecules by biochemical reactions. These reactions normally occur under anaerobic conditions.

Transformations, therefore, become the first biochemical reaction that affects your investigation. Your client developed a leak in his TCE storage tank. You cannot simply perform TCE analysis on the soil and ground water. You must include possible transformation products of TCE. You must look for the family of "dichloride" hydrocarbons and vinyl chloride. Looking at this another way, if you find other chlorinated hydrocarbons, you cannot assume that there is another source of contamination. The original TCE leak may be the source of all of the compounds.

This effect is not limited to chlorinated hydrocarbons. All organic compounds can be affected by natural biological activity. We must be aware of these possible reactions and broaden our investigation to ensure that these potential new compounds are included and their origins understood.

Soil Chemistry/Biochemistry

To truly understand a contaminated ground water, and the methods that must be used to clean it up, we must understand the environment in which the contamination occurs. In recent years, soil chemistry has become an important part of our investigations. We realize that the natural chemistry of the soil can interact with the contaminants in the vadose zone and the aquifer. The soil chemistry can also interact with the method that we select to clean up the site.

The natural chemistry of the soil is only part of the chemical reactions that can occur in the ground. Bacteria can add biochemical reactions to the environment. These biochemical reactions are not limited to transforming organic material as discussed previously. The bacteria can also affect anything from the valence state of the metals present to the gas content of the vadose zone.

To understand the environment of the soil and ground water, we must understand the chemistry and the biochemistry of the contaminated area. Without this understanding we will not have completed our investigation. The natural bacteria reactions are a part of the environment in which we find the contamination. We should make it part of our investigation.

Modeling

Modeling has become one of the most important tools that we use in an investigation. By the educated application of various models, we can project limited data to describe an entire contaminated site. This same model can then be used to help us analyze the proposed remediation method.

Contaminants do not move through the soil and aquifer at the same rate that the water moves through these areas. When we want to model the movement of organic and inorganic contaminants we include a retardation factor. The retardation factor describes the natural interaction between the contaminant and the soil. Most organic compounds are adsorbed to the soil. The strength of this adsorption and the rate of desorption varies for each compound. Therefore, we have to develop a separate retardation factor for each compound in order to accurately model the entire contamination plume.

To make the model really complete we must go beyond the effect of the soil interaction with the compounds, and we must include the interaction of the bacteria with the contaminants. Once again, the transformation process by the bacteria becomes important. In addition, the compounds can also be fully degraded by the bacteria. The model must include the disappearance of compounds, not simply the retardation of compounds. The model must also contend with the other side of transformation and include the appearance of new organic compounds. These new compounds will have their own unique chemical properties. If we do not include the biochemical reactions, we will not have a full understanding of the site.

Risk Assessment

The final area that I will cover under investigation is risk assessment. I put risk assessment under investigations because one of the most important factors in a remediation design is "what is clean"? Until we know the final objective of the remediation, we cannot design the optimum method. Risk assessment has been a major tool in determining the proper cleanup levels.

We tend to think of risk assessment as developing a single number that describes the concentration that will make a soil or aquifer safe. Bacteria can also have an effect in this area. Natural bacteria will degrade the organic compounds found in the soils and aquifer. Due to limitations caused by the lack of nutrients, oxygen, etc., the reaction rate of this natural degradation is slow. However, when we are addressing low concentrations of organics, as we do at the end of the project, then the rate of natural degradation can be significant. Therefore, an organic should not be described as a single number, but should be described as a number and a rate at which that compound will continue to naturally degrade.

Risk assessment can then be used to describe the goal of the project. However, this goal must be described as a combination of treatment to a specified level and then control and monitoring until the natural biodegradation finishes the project. I originally referred to this concept in my first article for *Ground Water Monitoring Review* (see Chapter 1). This process can be important in the selection of a final remediation. A quick example would be a remediation comparison between biological remediation of a soil and incineration of the soil. Assume that the bioremediation can reduce the toluene in the soil to 1 ppm in a three-month period. Incineration can reduce the toluene to nondetectable in less than two weeks. While the incineration will completely remove the toluene, it will also completely destroy the soil. The resulting ash will have to be buried in a landfill. The bioremediation will leave the soil intact and available for other uses once the cleanup is complete, but it will not completely remove the toluene. This view is not entirely correct. The proper way to describe the bioremediation is that the toluene concentration will be reduced to 1 ppm in three months by the bioremediation, and then the natural bacterial reactions will continue to reduce the toluene. Assume that the reaction rate is at a level that will half the concentration every six months. The remediation program can then be bioremediation with control and monitoring until the toluene concentration is at a level that the soil can be reused.

As can be seen by these four areas, bacteria can affect the investigation and description of a contaminated site. Unless we include these effects, we do not have a complete understanding of the site. The full description is necessary to design the optimum remediation program. The bacteria can also have a direct effect on that remediation.

REMEDIATION

We all view bacteria as part of an in situ program for remediation of organic contaminants, but this is only one way in which bacteria can be directly applied to remediation. There are several direct applications of bacteria and other indirect effects that bacteria can have on a project. Under direct use of bacteria, we will quickly review in situ remediation, including soils, vadose zone and aquifers, and we will review aboveground methods, including soil reactors, slurry reactors, and liquid reactors. Under indirect effect of bacteria we will review Vapor Extraction Systems (VES) and metal removal.

In Situ Bioremediation

Most of the readers of this book come from a geology, hydrogeology, or soils background. When thinking about an in situ biological remediation, your thoughts turn toward specialized bacteria (fu fu dust), clogging (well screens and soil pores), and magic (how the h... does this really work?) The real problems in an in situ cleanup relate more to geology, hydrogeology, and soil chemistry.

In more than 90% of all bioremediations, the natural bacteria will be the best bacteria to use for the cleanup. This even includes cases in which I have worked on creosote type contaminants including Poly Aromatic Hydrocarbons (PAHs). The main reason that the bacteria were not remediating the site on their own, was that they were limited by the lack of nutrients and oxygen. There will always be cases in which specialized bacteria will be able to contribute to a project (pesticides, herbicides, when the spill site is toxic to bacteria, and others). However, even in these cases, the rate limitations will be due to the application of nutrients and oxygen (for aerobic treatment).

All field applications of in situ biological remediation have been aerobic to date. There is some fascinating work going on in several laboratories on the application of anaerobic bacteria, but no successful field applications have occurred to my knowledge. We will, therefore, limit our discussion to the application of aerobic bacteria.

In situ bioremediation applications can be broken down into three main zones: the soil at the surface, the vadose zone, and the aquifer. In all three zones we must create an environment that optimizes the growth rate of the bacteria. The main medium of changing the environment and supplying the bacteria with nutrients is water. In the vadose zone and the aquifer, the water is also a major method for transporting the oxygen to the bacteria.

In general, the bacteria are present along with the organics in the contamination zone. The main design criteria is movement of the water to the zone, and control of the water after it passes through the zone so that it does not create a secondary plume. If these are the main design criteria, then hydrogeology is more important to the full-scale application of bioremediation than microbiology. The real problems that have to be solved are hydrogeological and soil chemistry. What type of wells or infiltration galleries are required to deliver the water to the zone of contamination? Will the nutrients and/or oxygen interact with the soils and prevent availability for the bacteria? What types of wells or trenches are required to capture the plume created by the water flowing through the contamination zone? A simple study on core samples from the zone can confirm the presence of the correct bacteria. There is no real microbiological design if the bacteria are there.

The main areas of in situ cleanup require different specific controls, but the basics remain the same. Surface in situ work (sometimes referred to as land farming), requires that the oxygen be supplied by exposing the soil to the atmosphere. Usually some type of tilling or disking is used. Nutrients are applied in a liquid or dry form to the surface. There is no need for wells or infiltration galleries to deliver these components, but we still must protect the site from secondary plumes from water application and rain. We must engineer leachate and run-off control measures as part of the project. Usually, I will not design a land farming operation unless the vadose zone and the aquifer below the soil are both contaminated. When I do, I use full liners, leachate and run-off collection, with storage and treatment of the water.

Vadose zone remediation requires that the water move through the contaminated zone to provide the nutrients and oxygen. Water is the only viable method to carry the nutrients to the bacteria. Air movement through the zone has been used for oxygen requirements. We will discuss air movement through the vadose zone in the next section. Once again, we must test to see if the bacteria are present and if the nutrients and oxygen will interact with the soil before they reach the bacteria. The main problem remains the correct design of a water distribution and collection system. Because of past problems with water collection in a vadose zone, I do not like to design in situ projects in vadose zones using water movement unless the ground water is also contaminated.

Finally, the aquifer requires the same design considerations. When we apply in situ bioremediation to an aquifer, it is because we need to treat the adsorbed material. The adsorbed material is slowly releasing to the water in the aquifer and forming the plume. We must eliminate the source of the contamination if we are ever to clean the site. The bacteria are probably within the soil matrix along with the contamination. Once again, simple tests can be run to confirm their presence. The real design work on the project is the injection and collection methods that provide a water flow pattern across the contaminated area. Then, in that water we place the nutrients and the oxygen. Laboratory tests are used to ensure that the nutrients and oxygen do not interact with the soil in the aquifer.

ABOVEGROUND REACTORS

Bacteria have also been applied to remediation in aboveground reactors. With aboveground reactors, engineering and operation of the reactor becomes the key to proper treatment. Once again, the microbiology is the easy part of the project. Ground water can be recovered and sent to a biological reactor. We can use any of the standard activated sludge or fixed film reactors. However, most groundwater flows are small (less than 100 gpm), and the organics are low in concentration (less than 50 ppm). Personally, I have been using some special submerged fixed film designs to resolve some of the unique aspects of groundwater treatment. No matter what design we employ, the basic requirements are the same. We must have a system that maintains the bacteria in the reaction tank, and we must supply oxygen to the bacteria. Nutrients are easily added to aboveground reactors.

When dealing with the soil, we also have to solve engineering problems. The basics are the same: oxygen and nutrient supplied to the bacteria. Three main methods have been used for soil reactors: land farming-type reactors, soil slurries, and soil piles.

As discussed in the in situ section, land farming has switched from a pure in situ process to one in which we must employ several engineered apparatuses. Once we have placed a liner below the soil to control leachate, a leachate collection system, and a water storage system, we really have created a soils reactor. We can even add air emission control to the reactor for volatile compounds that are not degradable. In this reactor we must still maintain the correct environment for the bacteria. We must supply the bacteria with nutrients and oxygen, and we have to provide some type of mixing.

The next type of reactor that has been employed for soils is the soil slurry reactor. In this reactor the soil is combined with water to form a slurry. It is easy to maintain the correct environment in the slurry. Standard mixing techniques can be used to continually mix the slurry and supply oxygen. Nutrients can also be easily kept at optimum concentrations. The main problem with the soil slurry reactor is that after the biochemical reaction has been completed we must separate the water from the soil. Getting water to separate from clay is not the easiest thing to do in a short period of time for a large amount of soil. Of course, if the soil is mainly sand then the liquid/solids separation will be less difficult. While the slurry reactor is optimum for the biochemical reactions, the soil slurry system is difficult to run because of liquid/solids separation problems.

The final soil reactor type that has been used in the field is the soil pile reactor. In this reactor the soil is excavated from the ground and placed on a plastic liner. As the soil is piled on the liner, nutrients and bacteria (if needed) are mixed with the soil. Perforated pipe is placed within the pile at regular depths and intervals. Once the pile has been finished, then all of the pipes are connected to the vacuum side of a blower. A plastic cover is placed over the entire pile to control rain water and air flow patterns. The pile contains all of the nutrients and bacteria that are required to complete the biochemical reactions. The blower then sucks air through the pile in order to supply the oxygen. Once the pile has completed the biochemical reactions, the soil can then be placed where desired.

Indirect Effect of Bacteria on Remediation

Two examples of indirect effect of bacteria on remediation are VES and metals. One of the most important new technologies that has been applied to remediation is VES. VES has been shown in the field to be able to remove volatile organics from contaminated soils. There are many technical papers that have detailed the operation of VES. However, few have run complete mass balances on the projects. Gasoline and diesel spills have been of particular interest. As mentioned in Chapter 2, not all of the organic compounds in gasoline and few of the compounds in diesel are volatile. However, VES has been reported to completely remove all of the organic compounds found in the soils.

This mystery has a simple answer. VES is similar to soil pile systems. While the nutrients will not be present, the bacteria will still be stimulated by the air movement through the soil supplying oxygen. Biochemical reactions are important to the complete cleanup of VES. The papers that have done material balances during their studies have shown this to be true.

My final example is metals. Most of us who have at least a couple of years experience in the field have encountered iron bacteria. Natural bacteria in the ground can use the iron in the water and soils as an energy source. Their natural activity will promote their growth on the slots of the well screen. As the bacteria grow they form a slime layer on the screen and slowly clog the slots. This eventually stops the flow into the well; some type of cleaning must be employed. The same effect can occur in packed tower air strippers. In this case the stripper loses its ability to treat the water and the packing must be cleaned.

Other metals can be affected by biochemical reactions. Usually the effect is limited to the valence state of the metal. I once worked on an aboveground treatment system that included biological treatment and arsenic removal. The natural state of the arsenic in the ground water made it unaffected by iron coprecipitation (the normal method of arsenic removal). However, we found that after the ground water passed through the biological treatment system, the valence state of the arsenic had been changed, and that it could then be removed by iron coprecipitation.

As can be seen by all of these examples, biochemical reactions have a large effect on investigations and remediations. In order to completely understand the contamination site and how to remediate it, we must include biochemical reactions. The reactions are simple when considered on an individual basis. As with everything new, the more we work with the bacteria the more comfortable we will be.

SECTION III

Developing Treatability Data

One of the strongest tools that we have to help us design groundwater and soil treatment systems is the laboratory. We can take a small volume of water from the aquifer or small quantity of soil from the ground and perform quick, flexible tests to determine the nature of the contaminants and the interaction of the contaminants with the soils of the aquifer and vadose zone. We can go a step further and test some of the methods for contaminant removal from water and soil.

While the laboratory can be of great assistance, it can also be the source of misleading information. We have to remember that the laboratory will always produce numbers. The problem is that it takes training and experience to be able to use those numbers in the right context. In other words, it also takes training and experience to design the experiments performed in the laboratory if the data is going to have any meaning.

This section will review what the laboratory can do for a design project. I have used a real-life example for the basis of Chapter 5. This chapter reviews the use of the laboratory to test physical/chemical methods for treatment of a ground water.

The other main area that abuses laboratory data is biological treatment. Chapters 4 and 8 and all of Section VIII cover the application of biological treatment to soil and ground water. In those chapters I have given some positive examples of using the laboratory to answer important questions for the design of a biological treatment system. Too often, however, the laboratory produces very expensive data and subsequent reports that do little to further our understanding of biochemical reactions at a particular site.

I was recently told by the manager of remediations at a major specialty chemical company that they had stopped using biological treatment at their sites because the laboratory tests cost too much. That company had come to the conclusion that they could not save enough using biological treatment in the field to overcome the initial laboratory expenses.

The main problem is that we have become too mesmerized with identifying the microorganisms that are at a particular site or developing a specialized bacteria for a cleanup. These are expensive activities. While most of these tests have their place, they do not belong at 90% of the cleanups. There are many obstacles to using biological treatment at contaminated sites. It is counterproductive to add extensive laboratory identifications to hinder biological treatment applications.

Biological and physical/chemical cleanups need the laboratory to answer important questions. We must make sure that the laboratory is the proper place to answer the questions that we ask, and that the questions are the correct ones.

CHAPTER 5

Using Laboratory Data to Design a Full-Scale Treatment System

Evan K. Nyer

INTRODUCTION

I was recently asked to review a Request For Quotation (RFQ) for a laboratory treatability study. After reading through the RFQ, I decided that it would be a great basis for a general review on the proper use of laboratory data and studies.

Most consultants envision a laboratory study as a scale model of the full-scale treatment system. They feel that as long as the correct scale factor is employed, any full-scale unit can be simulated by a small laboratory unit. The data generated by the laboratory equipment can then be used directly to design a full-scale treatment system. This perception is wrong.

The laboratory cannot be used to directly simulate full-scale treatment units. Laboratory studies can only be used to develop information about

ISBN 0-87371-731-7
© 1992 by Lewis Publishers

Table 1. Groundwater Organic Contaminants

Contaminant	Concentration, μg/l
Carbon Tetrachloride	50
1,1-Dichloroethane	65
1,2-Dichloroethane	35
Methylene Chloride	160
1,1,1-Trichloroethane	95
1,1,2-Trichloroethane	70
Trichloroethylene	230

physical, chemical, and biochemical reactions. Most full-scale unit designs are based upon solving practical problems that occur when one is trying to apply these reactions in the field. A good full-scale design is derived from a combination of the desired reaction and the solution to the practical problems.

In addition, laboratory reactions do not always simulate field reactions. For this reason all reactions cannot be performed in the laboratory. Some reactions are very accurate. Metal precipitation would be one example of a reaction that can be accurately simulated in the laboratory. Other reactions will not perform the same in the laboratory as in the field. Air stripping is a good example of a technology that performs differently in the laboratory and the field.

The best way to explain both of these phenomena is to review the RFQ. The laboratory can be a powerful tool in the design of full-scale treatment systems. However, we first have to understand what the data actually means, and how it can (or cannot) be applied to a full-scale unit.

FULL-SCALE TREATMENT SYSTEM

The purpose of the laboratory study in the RFQ was to simulate a full-scale treatment system that was planned for a site. It was desired to generate some operational data before investing in the full-scale treatment system.

Table 1 summarizes the organic compounds detected in the water at the site. They also expected to have about 5 to 10 mg/L of iron in the water. Figure 1 summarizes the treatment system that they wanted to employ to treat the water. As can be seen in Figure 1, the unit operations are:

1. Iron oxidation by aeration
2. Suspended solids removal by sand filtration
3. Volatile organic removal by packed tower air stripping
4. General organic removal by activated carbon adsorption

For each unit operation a laboratory test was specified. Let us review each test and see how the data was supposed to be used.

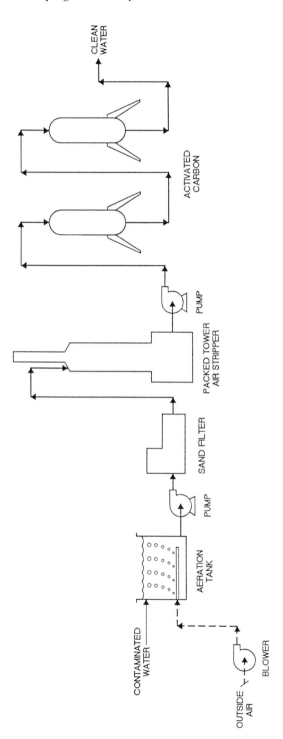

FIGURE 1. Water treatment system.

IRON OXIDATION

The full-scale oxidation system consisted of a reaction tank with a one hour residence time. A blower forced air through an air distribution system. The air provided the oxygen for oxidation and the mixing for the tank. Small bubble air diffusers were used to maximize oxygen transfer.

The laboratory test was supposed to be set up to simulate the full-scale system. A continuous flow tank with a one hour residence time was to be set up. Air flow and dissolved oxygen content were to be monitored to ensure that sufficient mixing and oxygen transfer would occur in the full-scale treatment system. Dissolved iron concentration at selected air flow rates was the main variable to be monitored.

Several problems exist with this testing procedure. First, air flow for mixing and oxygen transfer cannot be combined and scaled to a full-scale system. The scale factor for oxygen transfer is mainly affected by the depth at which the air is released in the tank. In general, this parameter has a linear relationship when extrapolating to a full-scale system. Mixing efficiency is related to the total power input into the tank and has an exponential relationship with the full-scale system. Neither of these critical design factors should be based upon laboratory data. Both should be based upon experience with full-scale designs. The laboratory test should only be used to test the effect of oxygen concentration on the iron oxidation rate. How you get that oxygen concentration is very different for laboratory units and full-scale units.

There is one other small problem with this test. Large water samples cannot be returned from the field without introducing oxygen. While we may pride ourselves in being able to take samples for volatile organic analysis without air space in the sample containers, we are not going to be able to perform that bit of magic with a 55-gallon drum. Even if there is no head space, there still will be small amounts of oxygen transferred into the sample. This oxygen will react with the iron in the sample. All of the iron will be oxidized before the start of the laboratory tests.

The only way to test the rate of iron oxidation on a groundwater sample is to perform the test in the field. The test must be run as soon as the water is taken from the well. Even in the field, great care should be taken to avoid the introduction of oxygen into the water sample when removed from the well. As a side note, I would suggest that pH be measured during the test. The pH has a dramatic effect on the rate of oxidation. The lower the pH, the slower the rate of reaction.

FILTRATION

Now that we have taken our soluble iron and turned it into insoluble iron (suspended solids), we only have to remove the solids in order to remove the iron from the water. The present system planned to use a sand filter for the solids/liquid separation step of the treatment system and the laboratory study was to simulate a sand filter.

The prescribed method in the RFQ was to set up a column with sand and to test the effect of various flow rates on the removal of suspended solids and iron. There were detailed specifications to ensure that the test was accurate. For example, the column had to be wide enough to ensure that there were minimal wall effects, i.e., thirty times the size of the sand particles.

The details were very interesting, but had very little to do with the operation of a sand filter. Sand filters do not strain suspended solids from water like filter paper does. Sand filters, and all particle filters, remove the suspended solids by solid/solid contact. The suspended solids come into contact with the sand particle and attach. The process has a closer relation to flocculation than to straining of suspended solids.

The only time that a sand filter acts as a strainer is when the suspended solids have filled all of the void spaces between the sand particles. At this point, the pressure required to force the water through the bed is too high and the filter is usually backwashed. This removes the suspended solids and opens the pore spaces once again.

The efficiency of the sand filter is related to the stability of the suspended solids. If the filter is not removing the required percentage of the suspended solids, then the chemistry of the water is changed, not the water velocity. A flocculent is usually added to the influent of the sand filter to remove more suspended solids. These results cannot be simulated in a laboratory-scale unit. And, even if these tests could be performed, they would not simulate the actual operation of the sand filter.

The main problem with the operation of the actual sand filter is cleaning it. Once the suspended solids are separated by the filter, can they be removed from the sand particles again? Sand filters usually fail because they cannot be cleaned, not because they are not removing suspended solids.

Even though the laboratory-scale unit looks like the full-scale unit, it cannot perform the same. Even with scale factors, the laboratory unit cannot simulate or predict the operation of a full-scale sand filter.

If it is necessary to remove a certain size or type of suspended solid in a treatment system, then a sand filter is the wrong unit of operation. In those circumstances the designer should rely upon a different type of filter, i.e., cartridge filter, ultrafilter, etc.

PACKED TOWER AIR STRIPPER

A full-scale packed tower air stripper cannot be simulated in the laboratory. The key to this type of air stripper is the performance of the packing.

The normal packing used in a tower is between 2 to 3 in diameter. Unfortunately, we cannot use smaller media and a scale factor, and we cannot afford to process sufficient water for media of this size in a laboratory setting.

The contractor must have realized this and did not provide specifications on the air stripper test. The RFQ simply stated that the air stripper should be simulated in the laboratory.

The only way to test an air stripper is to perform a pilot study in the field. The tower should be a minimum of 8 to 12 in diameter and 7 to 10 ft of packing depth. The testing should include running the air stripper from 10 to 30 gpm/ft². For an 8 in pilot tower, a range of 3.5 to 10.5 gpm would be required to fully test the packing. In addition, several runs of 30 to 60 min duration would be required to develop the data necessary for the full-scale design. As can be seen, a significant amount of water is required to test this technology.

There is one other method to acquire the necessary data. Most companies that sell air strippers have developed computer models that simulate the air stripper's performance. In addition, commercial computer models are also available for this purpose. As long as an accurate Henry's Law constant is available, then these models can provide the required data on air stripper performance in a particular situation. If the input data is not available, then a pilot plant is the only way to develop that performance data.

ACTIVATED CARBON

The specifications for the laboratory activated carbon unit were 30 gpm/ft² and a 30 min residence time. When I first received the RFQ, I thought that this was a typographical error. A 120 ft tall column would be necessary to meet these specifications. You may perform your own calculations to confirm this.

After calling the author of the specifications, I found out that they were simply following the specifications that were listed with the full-scale carbon unit. The problem is that the 30 gpm/ft² refers to the design limitations of the tank and piping, not the carbon. The 30 min residence time refers directly to the carbon itself. The RFQ combined these two into one specification for the carbon. After some detailed discussion, it was decided to use 3 gpm/ft² as the basis of the carbon system. This would be a normal flow rate for a field carbon unit.

This left one problem with the laboratory test. We calculated the amount of contaminated water necessary to break through the activated carbon in the laboratory unit. This can be done by using the isotherm data provided in Chapter 3. We cannot measure the performance of the carbon column until the compounds start to come through the unit and can be measured. We estimated that 5000 to 6000 gal of contaminated water would be required for the test.

I would like to know if anyone has ever approached a government agency and told them that they were going to remove 5000 gal of untreated water from a superfund site and send it to a laboratory. Of course, this is too much water to bring into a laboratory. The other thing to remember is that all of this water would have to be pretreated with the rest of the treatment system in order to perform an accurate simulation, and the water would have to be disposed of in an appropriate manner.

There are three choices when performing tests on activated carbon. First, isotherms can be performed in the laboratory. This data is easy to obtain, but is not always accurate when comparing it to activated carbon in a column. The second choice is to perform a field pilot test with a small column of activated carbon. The data is very accurate, but the test can get complicated if pretreatment is required. The third choice is to send a sample to Calgon Carbon. They have developed a laboratory test using powdered activated carbon in a column that simulates full-scale units.

There is one other method to get the required data. Most of the companies that sell activated carbon have developed computer models that simulate the performance of carbon in a column. These computer models usually work for individual compounds or mixtures of organic compounds. These models can be used to design most of the carbon applications for the field. The carbon companies will also know when the models should not be used and field data is required. The models are probably more accurate than any laboratory test. The laboratory should be used to test for compounds and conditions that will interrupt the performance of the carbon. I have seen more carbon columns ruined by suspended solids than by the organic content of the water.

GENERAL SPECIFICATIONS

The last specification of the RFQ was the requirement that all of the sample had to be stored at 4°C while it was waiting for laboratory processing. In general this is a good idea, but we should evaluate whether the practical problems this may cause do not outweigh the technical results. The main benefit that the low temperature provides is to ensure that there is no biological activity while the sample is being stored. Because the primary organic compounds in this study are not degradable, the low temperature storage will not gain that much accuracy. On the other hand, storing 5000 to 6000 gal of water at 4°C will be difficult.

SUMMARY

The laboratory provides a powerful tool that can be used to derive costs and answer technical questions concerning water treatment design. The problems described in this chapter come from the misconception that all full-scale processes can be simulated by laboratory scale treatment units. This is simply not true. We can always generate data from a laboratory unit, but we must make sure that the data is applicable.

When starting to design a treatment system there is always a requirement for data to be used as a design basis for the unit operations. We need to define exactly what data is required and then decide what is the best way to generate that data. We need to use a combination of laboratory studies, pilot plants, literature, and experience in order to develop an accurate and productive design.

SECTION IV

Developing Detailed Designs

This section does not need a long introduction. Every remediation project requires a team of specialists to develop the data and to design the remediation. Some of the technical experts that are required on a project are: hydrogeologist, geologists, geochemists, modelers, risk assessment experts, engineers, construction managers, operators, etc.

I wrote this chapter to give the readers a better understanding of what role engineers contribute to a remediation project and how they communicate their efforts. The National Ground Water Association is mainly made up of geologists and hydrogeologists. Therefore, the main readers of *Ground Water Monitoring Review* are from the same background. This chapter was written in response to comments that I had been hearing between these two expertise: hydrogeologists and engineers. There seemed to be a basic misunderstanding by each group of what the other group contributed. Of course, this is a nice way of stating the misunderstanding. The actual discussions degraded to the level of training at certain colleges and, eventually, to thought-provoking discussions about personal lineage.

The important point of this section is that a good project manager realizes that his or her job is not to do everything on a project. A good manager assembles their team of experts and relies on their training to perform individual tasks. The manager coordinates between the experts getting the remediation done on time and under budget. (I'm just kidding about the budget.)

CHAPTER 6

Engineering

Evan K. Nyer, Peter Boutros, and David Keyser

Most of my readers are geologists and hydrogeologists. For years, the main endeavor of these professionals has been investigation of contaminated sites. We are all now entering a stage in which many of these investigations are going to the next phase, remediation. As we start to remediate the sites we find there are several new disciplines that are required to complete the job. One of the main disciplines is engineering. The idea of this article is to delineate many of the specific tasks that engineers perform, and to explain the reasons behind those tasks. This should provide a greater understanding of the engineer's task on the remediation project.

VENDOR SELECTOR VERSUS DESIGNER

We first must understand the difference between a Vendor and a Designer. Most of us have had experience in a groundwater situation that was small

ISBN 0-87371-731-7

flow and required simple treatment. Many of the groundwater contaminants are volatile organic compounds, and treatment can be satisfied by air strippers or activated carbon. It doesn't take four years of college in an engineering discipline to be able to call up an air stripper supplier. They can provide the details of the design, the proper sizing, the installation requirements, and anything else that is required for solving the treatment problem. The first question becomes the need for engineers on a small project like an air stripper for a gasoline station remediation.

Many remediations are going beyond the simple treatment for a small gas station spill. We are now encountering projects that require large flow rates and new, multiple technologies for treatment of the ground water. Does the treatment formula really change? We are still able to call up vendors that supply these technologies, get them to supply the detailed design, drawings, etc. Do we need a four-year or advanced degree in engineering in order to make phone calls and gather the information from vendors?

The answer to both of these questions is yes, we do need engineers on both types of projects. First, there are the legal reasons that engineers are required, i.e., signature of drawing submittals that are required by city and state organizations, state requirements for engineering stamps on any construction, etc. But the legal implications aren't the main reasons. The main reason is that engineers perform process design.

The difference between selecting a vendor and process design is an understanding of the technology and the methods of implementing that technology. I was once asked to review a design for a groundwater treatment system. The system included 20,000 lb activated carbon units. I called up the designer and asked what was the basis for the selection of 20,000 lb activated carbon units. The answer was they they had called Calgon Carbon Corporation, and Calgon had specified 20,000 lb carbon units. I asked the designer if they realized that they just based their design on the size of Calgon's trucks. Calgon uses 20,000 lb carbon units because that is the maximum amount of carbon that can fit on their trucks. This is not to say that the size of Calgon's carbon trucks are not a good basis for design of an activated carbon unit. However, the process designer needs to understand the basis of the design not just the vendor selection. The vendor's design basis may not be optimum for a particular situation.

This does not mean that process engineering can be accomplished simply by having an engineer call up the vendors. I see this too often, and I consider it "lazy" engineering. Process engineering must be performed by someone with experience and understanding of the technologies involved.

Process engineering is the backbone of any remediation design. The most critical point encountered during a project is the technology selection by the process engineer. It is actually the system architecture on which the implementation strategy of the project will depend. Available technologies have to be evaluated and only the most effective technique should be considered for the particular application. We do not want to start applying technologies based on the vendor being a good salesman. Without the engineer's experience and knowledge a project may not provide the remedial treatment intended. This includes small projects. Do we believe a vendor when he tells us that cleaning the media on his air stripper is simple when dealing with small flows? He has not lied. But, the whole truth is that different technologies may be a better selection for small flows with iron present. The vendor did not have that technology to offer. It is the process engineer who is responsible for understanding all of the available technology and making a fair and complete comparison.

Even after we decide to use a specific vendor's technology, we still need the process engineer. When a specific equipment is produced by a manufacturer it is usually produced in a few different sizes to meet the bulk of the demands. Sometimes the most efficient and useful unit for the application falls between sizes and cannot be used as manufactured. The engineer's job is to make the decision on using the piece of equipment as is, modify it, or go to a different manufacturer. It can be a costly mistake over the long run if the equipment is supplied and installed just because it was available or because it was low-priced.

Vendors will probably supply the right system on 90% of the simple jobs and only 30 to 50% of the complex jobs. Process engineering increases the odds for a successful project. Do you want to go to the client and inform him that you saved $10,000 but the system does not work?

VENDOR SPECIFICATIONS VERSUS DETAILED DESIGN

The second area that requires an engineer is the detailed engineering. First, every unit operation is made up of many details. Everything from the thickness of the steel to the type of paint must be selected. Even when we decide to recommend a vendor's piece of equipment, we must review the detailed design to ensure that the vendor's standard selection will meet the requirements of the specific installation. In addition, every project that includes unit operations for treatment, and some type of movement of air or water, will require tanks, pipes, electricity, etc. Once again, it is simple to look in the yellow pages and select a contractor to make the detailed selections for a specific location. However, if I am a project manager on a remediation project, I am not about to let a contractor make those important selections that can cause a project to succeed or fail. I am the one that has to go to the client if it fails, not the contractor. Detailed engineering, under the control of the project manager, provides the specifications that are required to ensure that the small parts of the treatment system work.

Something as simple as the packing material of an air stripper can cause inefficiency of the system. Most vendors use a standard packing with their air strippers. Once again, the local ground water may contain high levels of iron. The vendor has never installed a unit in this area and does not realize that the standard packing may cause problems. The detailed design by the engineer would include a review of the different packings and selecting the most appropriate type.

Another project that we are working on would serve as a good example for detailed engineering requirements. The opposite of a planned and engineered project is one that is "designed-as-you-go". The decisions made for this type project are made in the field and may or may not have limited documentation that is given to the construction contractor. In this particular example we had a case where eight wells were manifolded and then piped to a treatment system. No provisions were made to monitor or control the individual flows from each well. Also, no provisions were made for instrumentation. It was assumed that since each well pump could produce 200 gpm, that the system would produce 1600 gpm with each well producing an equal amount. This is a big assumption. A balanced flow and a controllable piping system is very important to the treatment system and, even more so, to the groundwater plume which must be pumped from as directed by the hydrogeologist.

The problem is that there is no way that the wells simply piped together will produce equal flow rates. The more resistance a pump has to push against, the lower the flow rate that it will produce. Length of pipe, elbows, valves, etc. all contribute to resistance of flow in the pipe. Basically, the wells closest to the treatment system will have less resistance and more flow rate. The only way that a multiple well system can have a controlled flow rate is by detailed engineering. The engineering would include an understanding of the head losses in the pipe, and a specification of control and measurement techniques for the full-scale system.

Since this particular system is already installed, there is no way for us to go back and put the controls on the existing buried pipe. We also have no way of informing the hydrogeologists of the flow from a specific well. Most important, the hydrogeologist cannot change the flow pattern in the well field to optimize the groundwater flow for plume control. The project saved money and time by skipping detailed engineering, but at what cost?

COMMUNICATION

Process design and detailed design are obvious jobs for the engineer to do. If we work with good engineers, we develop a respect for their ability to perform these two functions. With that respect, comes a belief that the engineer should be part of the remediation project. At a minimum, on simple projects, I would suggest that an engineer review the designs of a vendor or contractor before installation begins.

Finally, we come to the question of what do we actually get from an engineer? Obviously, the engineer provides the reports, specifications, and drawings required by the regulatory agencies. Likewise, these documents are used to obtain permits for construction and finally to provide a record of the entire construction and installation. Most geologists and hydrogeologists accept reports from engineers (with minor rewriting). The item they have the least respect for is engineering drawings. Typical feelings that I have heard about engineering drawings range from confusing to a total waste of time. You must understand that an engineering drawing is simply a form of communication. Like the old proverb, "a picture is worth a thousand words", the engineering drawing provides the details in picture form. Following are some of the main types of engineering drawings and their specific uses.

The Process and Instrumentation Diagram — Figure 1 describes the flow of the system and shows the controls required for its efficiency. This includes all the instrumentation and their interaction as well as the major manual controls. The instrumentation symbolism and identification techniques shown on this drawing constitute a precise language which communicates concepts, facts, and instructions. These details range from very simple to very complex according to the process requirements and the intensity of the instrumentation. Each call-out or "balloon" has letters and abbreviations which refer to industry standard instruments and hardware. This drawing, therefore, becomes the major link between process engineering and detailed engineering.

The Control Panel Layout — Figure 2 shows the layout of the enclosure and the inside panel of the control cabinet. Coupled with a bill of material, this drawing specifies every component and assigns its position in the assembly. To simplify the operator task, the layout of the push buttons is designed to be in sequence with the process flow logic. The call-outs refer to a bill of materials (not shown) and the actual layout dimensions could be given on an additional detail sheet. Proper wiring procedures and ease of installation require a layout such as shown in this drawing.

The Piping Layout Isometric — Figure 3 is used to communicate piping information for visualizing the system and estimating the construction costs. Additional scaled two-dimensional drawings with lengths and elevations could be supplied to give further detail if required. Or, a highly detailed, computer-aided three-dimensional drawing could be created if the size and complexity of the project justified the detail.

The example drawing shows basic equipment relationships and gives specific piping information called out in the "balloons" or call-outs. Fittings are called out by their numbers which are referenced to a separate materials list or bill of materials (not shown). Valves and pumps are further represented by a "V" and "P" prefix, which are also specified on the bill of materials. Likewise, a "U" prefix refers to a chart which specifies pipe labeling and painting.

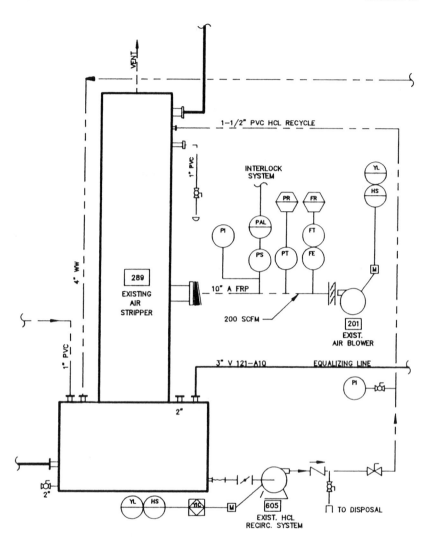

FIGURE 1. Process and instrumentation diagram.

Notes added to the drawing give specific detailed information where required and pipe sizing is indicated directly for quick reference. Finally, the type of symbols used for the pipe fittings are standardized to represent either a welded, flanged, or threaded connection. All the information on such a drawing is clear and concise. This allows the engineer, the client, the contractor, the inspectors, etc., to efficiently communicate and understand the job as designed.

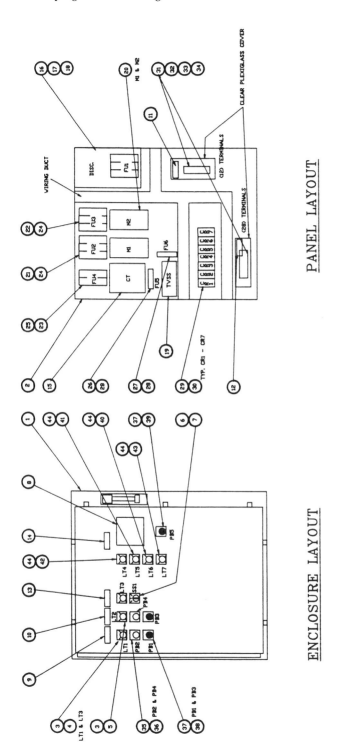

FIGURE 2. Control panel layout.

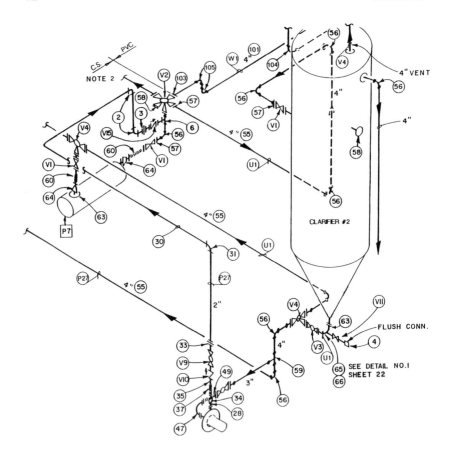

FIGURE 3. Piping layout isometric.

Remediations will require engineering experience, knowledge, and detailed communication from the engineers to be a part of a successful project. The project manager needs to understand how to integrate this discipline into the project team.

SECTION V

Unit Operations

Unit operations are the backbone of any general discussion on the treatment of ground water or soil. This is the section where we discuss, in detail, how the contaminants are removed from water and soil. Most authors would spend most of their efforts on this subject matter and little time on peripheral areas. However, as the reader will notice in this book and other publications I have written, I consider this area important, but not any more than one of several important areas in remediation. With groundwater treatment it is not just the selected technology, but rather how that technology is applied.

Groundwater treatment is unique. Never before have we tried to clean a body of water. In the past we have treated and removed the sources of contamination entering the body of water. The river, lake, or ocean were left to clean themselves by natural processes. Ground water will clean itself, but we find that the natural rate is too slow. While the first step is to remove the source of contamination, we go beyond this effort with ground water and proceed to clean the water body (aquifer) itself.

This creates a design criteria that is different than any experienced before.

Most treatment experts were originally trained to treat wastewater. Most of the unit operations that we use, with the exception of packed tower air strippers, were originally designed for treating wastewater. While this training and all of these technologies are applicable to ground water, we cannot simply directly transfer this knowledge. We have to understand the essence of the technology and then adapt it for use with the unique requirement of cleaning a body of water. For example, biological treatment can be applied to ground water, but the activated sludge process is probably the wrong way to do this. Each unit operation will have this limitation. Each must be applied with the unique requirements of ground water, not simply based upon the ability of the technology to remove the compound from water. The same basic message holds true for soil treatment.

This section covers three unit operations in detail. Each chapter shows how to apply these technologies to ground water. The rest of the book has examples of unit operations being applied to remediations. Other books and articles can be found that discuss treatment technology and unit operations. This book concentrates primarily on how to apply these technologies.

CHAPTER 7

Would the Real Air Stripper Please Stand Up?

Evan K. Nyer and Jodie Montgomery

Everyone is somewhat familiar with air stripping as a method to remove volatile compounds from ground water. In this chapter, we will review some of the basic assumptions that we have been using while applying air stripping technology.

Air strippers have established themselves as the most popular groundwater treatment method. There are two main reasons for this. First, most of the compounds that are currently being treated are volatile compounds. Gasoline stations and chlorinated solvent spill sites represent most of the current groundwater remediation activity. The benzene, toluene, xylene, and ethyl benzene from gasoline and the methylene chloride, trichloroethene, carbon tetrachloride, and various other low-molecular-weight chlorinated hydrocarbons can all be reduced to the low ppb (<5 ppb) range by air stripping technology.

The second reason for air stripper popularity is cost — air strippers are cheap. Construction and operation of an air stripping tower costs very little.

ISBN 0-87371-731-7

When the compounds are volatile, there is no less expensive way to remove them from water. I will even go as far as to say that there will never be a less expensive method of removing volatile compounds from water. What would be less expensive than a pipe, some plastic media, and a low-horsepower blower?

Actually there are three reasons for the popularity of air strippers. The third reason is that air stripping can be modeled. Not only modeled, but the model involves simulations that are complicated enough that a computer is required to do the calculations. What engineer could resist the lure of "designing" an air stripper? Of course, being a nice guy and an engineer, I've left it at two reasons.

Air strippers are so popular that people keep forgetting to design them. They simply assume that air stripping towers are the right technology, plug the numbers into the computer, and "voila" — groundwater treatment system.

Well, I hate to be the one to ruin the party, but this mentality results in the specification of strippers that are not practical in the field. There are several other factors that should be included in stripper design so that major adjustments are not required in the field. We seem to specify an air stripping tower not matter what the situation, and simply add on to the design in order to make it work.

Regulatory requirements limit volatile organic emissions from the stripper, so we add an air treatment unit. Air stripping operations can be significantly controlled by federal, state, and local regulations on air emissions. Air regulations often limit the amount of organics that may be released to the atmosphere.

Compounds in the water do not strip readily, so we simply add heat. Physical and chemical factors for the "volatile" compounds may limit removal efficiencies. Certain volatile compounds — acetone, methyl ethyl ketone (MEK), etc. — are highly soluble in water and are not effectively removed by a standard air stripper.

Iron in the water is fouling the media, so we simply add a pretreatment system. Iron, hardness, bacteria, and other nonhazardous components of the ground water can ruin the operation of the tower. Often, soluble constituents, such as iron, will form insoluble oxides during aeration, which will accumulate in the air stripper. This fouling will eventually interfere with stripper operations.

In all of these cases we never give up our precious air stripping tower, we just put Band-Aids® on the system. We have got to get out of the mindset that air strippers are a cure-all! There are other methods of designing air strippers and there are other technologies for treating volatile compounds.

All of these problems can be avoided if designers will take a more practical approach to air stripper design. Instead of simply attaching extra equipment to the air stripper, we need to start thinking through the entire design from scratch. The worst example of "blind" design that I have seen, to date, is a design that included a $500,000 pretreatment system to remove iron for an existing $20,000, 15-gallon-per-minute (gpm) air stripper.

We will not completely review air stripping design methodology in this chapter. But I want to try to open your mind a little. Let me concentrate on one area. Packed towers are only one way to design an air stripper, so let us compare the use of packed towers with two other feasible alternate strippers: cooling towers and diffused aeration tanks.

STANDARD AIR STRIPPER

The best place to start with this comparison is to review the specific design of an air stripping tower. One of the problems with current design thinking is that we envision air strippers and the packed tower design of an air stripper as the same thing. Packed towers are only one way of stripping volatile compounds from water. Taken from my book, *Ground Water Treatment Technology*, the following is a detailed description of a packed tower:

The major components of a stripping tower are the tower shell, tower internals, packing, and air delivery systems (Figure 1). The tower shell is usually cylindrical, for strength, for ease of fabrication, and to avoid any corners that might induce channeling of the air or water.

Materials of construction include aluminum, stainless steel, coated carbon steel, and fiberglass. Fiberglass and aluminum are the least expensive, with stainless steel and coated carbon steel being slightly higher in cost for most cases.

The tower internals serve to ensure that the mass transfer process takes place under optimal conditions and at the most economical cost. Starting at the top of the tower, the first component that requires attention is the air exhaust ports. These ports are typically located around the circumference of the tower, and are sized to permit the air to escape with a minimum pressure drop. If the tower is for potable water, the outlets should be screened to prevent contamination by windborne material entering the tower; towers screened with 24-mesh screen have no reported problems in this regard.

Continuing downward, the next component encountered is the mist-eliminator system, placed in the tower to prevent the discharge of large quantities of water entrained in the airstream. This is accomplished by forcing the airstream through a series of bends to impinge the water droplets on the surface of the mist eliminator. When the droplets grow to a large enough size, they fall back into the tower.

Water is introduced into the tower by means of a distributor, which ensures that the water is evenly distributed across the surface of the packing, while

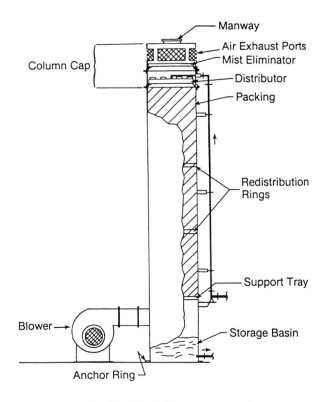

FIGURE 1. Packed tower components.

allowing for smooth, unimpeded airflow upward to the top of the tower. The distributors fall into four general categories: distributor trays, trough-and-weir arrangements, header-lateral piping, and spray nozzles.

Below the distributor lies the packing, which is held up by a packing support plate. This plate must be structurally capable of supporting not only the weight of the packing, but also the weight of any water present in the packed bed. At the same time, the plate must have enough open area to minimize flooding, a condition that results when the water flows downward through a tower and is significantly impeded by the gas flow.

The design of the base of the tower will vary with system configuration; an integral clearwell (storage basin) may be supplied as part of the tower, or the water may flow by gravity to discharge in a stream or sewer.

The single most important component selection is the tower packing. The ideal tower packing will provide a large surface area for the air and water to interact, and it will also create turbulence in the water stream to constantly expose fresh water surfaces to the air. The packing should have a large void area to minimize the pressure drop through the tower.

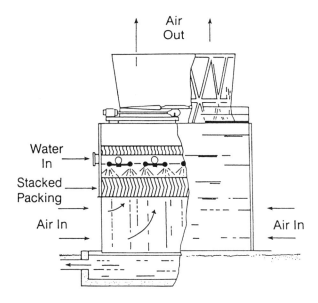

Air
Out

Water
In

Stacked
Packing

Air In

Air In

FIGURE 2. Induced draft counterflow tower.

The final component of an air stripping system is the air delivery system. Usually a forced draft blower is provided at the base of the tower or housed nearby in a building, if noise control is a design constraint.

ALTERNATIVE AIR STRIPPER DESIGNS

The packed tower previously described is the most efficient method of contacting air and water in order to achieve high percent removal of volatile compounds. However, there are other possible designs. These should be used when the conditions are appropriate. Let us look at two scenarios for which an alternative design may be better.

Scenario 1: Packed Towers Versus Cooling Towers

For high flow rate remedial designs (>1000 gpm), cooling towers may be more cost-effective and easier to operate and maintain than packed towers. The standard cooling tower is similar in design to the packed tower, although there are numerous configurations for cooling tower design based on the application. A typical configuration is the counterflow tower as shown in Figure 2. Air flow is induced using a fan at the top of the tower. The rest of the components are very similar to an air stripper.

One of the main differences between a cooling tower and a packed tower is that the cooling towers use different packing media. The packing is structured and is placed in the unit as large bricks of media. The water is usually distributed over a film sheeting to enhance air to water contact. When using a cooling tower for volatile organics removal, the air to water contact can be further optimized by removing the commonly used film sheets and spraying the water into the tower.

Consider the case in which approximately 1000 gpm of ground water containing 25 parts per billion (ppb) trichloroethylene requires treatment to reduce the trichloroethylene concentration to less than 5 ppb, approximately 80% removal. We contacted Duall Industries, Inc. in order to get a rough cost on a standard air stripping tower. It provided an estimate for an 8-ft diameter air stripper with 30 ft of packing. The system could be purchased for approximately $50,000 to $60,000. A 15 to 20 hp blower would be required to provide approximately 10,000 cfm air.

The TCE removal could also be accomplished by using a $20,000 packaged cooling tower, as specified by Marley Cooling Tower Company. Marley has a Model 2210 cooling tower that supplies approximately 106,600 cubic ft per minute (cfm) air to the 1000 gpm water using a 15 hp fan. The tower is approximately 9 ft in height and requires a 14 by 14-ft platform for placement. One added advantage to the Marley system is that this is a standard unit design and could be delivered in a couple of weeks.

Although cooling towers do not look like the typical air stripping system, they may be the most cost-effective solution. Many people may balk at not seeing an ''air stripper'' specified for the cleanup. My own experience with using a cooling tower to clean up ground water resulted in severe problems because the state representative felt that the unit in question was not an air stripper, because it ''was not round''.

Scenario 2: Packed Towers Versus Diffused Aeration Tanks

Another situation in which air strippers are frequently, blindly specified is for the removal of volatile organic compounds from streams containing high concentrations of inorganics. Let us consider the case in which 5 gpm of ground water must be treated to reduce 1 part per million (ppm) trichloroethylene to less than 5 ppb. Assume that the ground water also contains 25 ppm iron and that there are no effluent restrictions on the iron.

Several price quotes were obtained for a packed tower. The unit was designed for 5 gpm (it could treat a maximum flow rate of 25 gpm at the minimum 10-in diameter). To avoid wall effect, the minimum diameter of the tower was 10 in, with 2-in tripacs, and the packing height was 19 ft. The budget price was approximately $8000. This estimate includes the tower, blower, and internals.

It was assumed that due to the high iron content of the ground water, the packing would require replacement every two months. I can cite numerous cases, based on previous field experience, where packing replacement was required even more often. Replacement packing would cost approximately $350. Disposal of the used packing, and labor would cost about $450 each time the packing was replaced. The total packing replacement cost per year would be approximately $5000.

A diffused aeration unit could also be used to reduce the TCE concentrations and would not present the fouling problems typical to packed towers. Diffused aeration tanks are simple units consisting of a tank, a small bubble air diffuser, and a blower. This is such a simple design, I am not even going to bore you with a figure. The water simply enters the tank and is put into contact with air from a diffuser placed at the bottom of the tank. The volatiles transfer from the water to the bubbles and are removed from the system.

High air-to-water ratios must be maintained to meet the required treatment efficiency of 99.5%. A 5 gpm diffuser unit with a 2 to 3 hp blower and a tank sized for a 30-min retention time (300 gal) would cost approximately $5000. The capital cost of this system is lower than the cost for the packed tower. The operating costs are significantly lower even though we will require more power for the diffused aeration system. The difference in electrical costs between operating the packed tower and the diffuser unit is approximately $1000 per year. Even at higher flow rates, the media replacement costs will be more expensive than the difference in power requirements.

CONCLUSIONS

As you can see, air stripping for volatile organic removal can be accomplished with other system configurations besides the typical packed tower. Cooling towers are feasible alternatives when high air-to-water ratios are required and treatment efficiency requirements are less than the 80% removal range. Cooling towers are also easier to obtain than custom-designed packed towers.

Diffusers may prove to be a great alternative to using packed towers for low flow, ground water contaminated with inorganics. Additional monies may be expended in electrical costs but, at a minimum, a cost comparison should be performed to evaluate electrical costs versus packing replacement costs, or other pretreatment alternatives.

Don't put Band-Aids® on your stripper. Think through the entire process. Remember that there are other technologies out there. Alternative technologies may provide a more economical solution.

CHAPTER 8

The Application of Biological Treatment to a Landfill Leachate

Evan K. Nyer

The application of biological technology to contaminated ground water is probably the least understood of major technologies being used today. Many articles have been written on in situ biological treatment, but very few have been presented on above-ground biological treatment systems.

The case history presented in this chapter includes a landfill leachate contaminating a nearby stream, unusual compounds (toluic acids), laboratory treatability studies, and a full-scale installation that has been in operation for more than a year. The final point of interest is that this biological treatment system has been operating at a closed industrial site with a minimal amount of operator attention. Additionally, this is a project on which I have been personally involved.

ISBN 0-87371-731-7
© 1992 by Lewis Publishers

BACKGROUND

The site had been used by various industrial concerns since the early 1900s. A chemical company purchased the property in 1946. A landfill was started on the property in the late 1940s; the landfill was closed in 1973. It was reported that dry alkaline wastes, off-grade detergents and silicates, floor sweepings, fiber and steel drums, pallets, metal scrap, and miscellaneous wood, paper, etc., were landfilled during this period. Disposal of wastes from the manufacture of organic chemicals (such as pyridine-3-carboxylic acid, insect repellents, toluic acids, and others) was not formally mentioned but probably occurred with the startup of the organics plant in 1957.

The landfill site was backfilled and seeded in 1973 and capped with 1 foot of clay in 1979. Breakouts of brown leachate occurred both before and after closure. The normal operating procedure was to repair the leaking spot by adding clay. The most recent problem occurred in April 1986. Snow and rains had damaged the cap several days prior to an inspection. Recovery of both the landfill leachate and surface run-off waters was immediately undertaken.

As a first effort a simple earthen dam was constructed across the natural surface drainage from the site to contain the leachate. Large amounts of both the leachate and surface runoff were collected. Eventually, a gravel-filled subsurface collection system was installed to collect leachate.

The initial agreement with state environmental regulatory personnel involved transport of the leachate off-site to a treatment facility in New Jersey. Approximately 66,000 gallons of water were transferred and treated at a cost of $0.20 per gal, transportation included. The plant then requested that it be allowed to treat the leachate on site. Permission was received from the state to treat the leachate with activated carbon columns as long as a 5-pounds-per-day total organic carbon (TOC) discharge limit was met. The TOC limit was increased to a single-day 20-lb maximum and a 10-lb-per-day average at a later date.

The primary organic contaminants in the leachate are toluic acids. Concentrations of these chemicals were generally in the 300 to 400 ppm range. Figure 1 shows the chemical structures, melting points (MP), boiling points (BP), and solubilities for the three isomers present. In addition to toluic acids, isomeric xylenes, ortho-methylbenzyl alcohol, sulfur, and tetramethylbutyl-phenols were also detected in low concentrations. The compound with the highest concentration (other than the toluic acids) was ortho-methylbenzyl alcohol with a 1.66 ppm concentration.

LABORATORY TREATABILITY STUDIES

While activated carbon treatment was progressing, the property owner

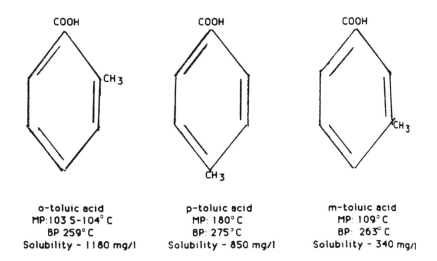

o-toluic acid	p-toluic acid	m-toluic acid
MP:103 5-104° C	MP: 180° C	MP: 109° C
BP 259° C	BP: 275° C	BP: 263° C
Solubility - 1180 mg/l	Solubility - 850 mg/l	Solubility - 340 mg/l

FIGURE 1. Chemical structures, melting points, boiling points, and solubilities for three isomers.

was interested in seeing if biological treatment could be used to remediate the leachate. Several laboratory studies were planned to investigate if biological treatment could successfully reduce toluic acid concentrations under aerobic or anoxic conditions, and to generate process design data. The results of the laboratory tests would also be presented to the state regulatory agency, which did not consider biological treatment to be a viable alternative treatment technology at this time.

Successful biological treatment of wastes require favorable environmental conditions, suitable microbial populations, and the absence of high concentrations of toxic/inhibitory chemicals. The laboratory experiments were designed to test these conditions.

The first phase in any biodegradability study is to review the technical literature. Past work by Alexander and Lustigman[1] showed that aerobic soil organisms are at least able to cleave the ring structure of toluic acids. Some anaerobic organisms are also able to reduce and cleave benzene rings.[3] Thus, it was possible that either aerobic or anaerobic organisms (in their respective environments) could be used to remediate toluic acid contamination.

The next phase in the treatability study was to determine if the landfill leachate was suitable for biological treatment. If the leachate was acutely toxic or inhibitory toward microbial life, biological treatment of the leachate would most likely have to be abandoned. There are many methods available for testing biological activity in the laboratory. I have found that simple microbiological techniques can usually be used to gather the necessary data.

The leachate was initially tested for microbial toxicity/growth inhibition. More expensive and complicated tests can be used to gather this information, but they should be reserved for more appropriate problems.

Microbial Toxicity/Growth Inhibition Tests

The protocol used for these studies is modified from a previously published procedure.[2] The modified test exposes an inoculum of microorganisms, inorganic nutrients, and organic nutrients to different concentrations of the leachate. Under favorable environmental conditions, cell growth causes an increase in the test solution turbidity. The amount of growth is quantitated by spectrophotometrically measuring the amount of light passing through each test solution. Duplicate tubes were used for all dilutions and controls. The positive control tubes consisted of 10 ml of distilled water augmented with 0.1 ml each of organisms, organic nutrients, and inorganic nutrients. The negative control was prepared in a similar fashion but contained mercuric chloride as a microbial poison. "Inhibited" growth was said to occur in samples whose increase in turbidity was less than that obtained with the positive control tubes. No growth inhibition is said to occur if the amount of growth in the test solution equals or exceeds that obtained in the positive control tubes.

The growth inhibition test was performed twice: once using a leachate at an unadjusted pH of 8.7 and again with the pH adjusted to 6.6. Results for duplicate samples showed excellent reproducibility. Growth was observed at all leachate concentrations at each pH. However, the leachate at a pH of 8.7 exhibited growth inhibition when present at 50, 75, and 100% strengths.

The fact that the leachate is inhibitory at certain concentrations does not preclude the use of biological treatment as a remediation technology. Rather, minor growth inhibition may merely reflect the fact that the microbial inoculum used is not totally acclimated to the chemicals present in the sample. Acute inhibition, such as that observed with the poisoned tubes, would be much more indicative of potential remediation problems. It may also be possible to dilute the waste stream with clean water prior to biological treatment.

Aerobic and Anoxic Biodegradability Tests

Since the leachate was not acutely inhibitory/toxic to microbial growth, further experiments were conducted to demonstrate the biodegradability of isomeric toluic acids. Because of the expense of quantitating the toluic acid isomers directly, biodegradation of the compounds was to be monitored indirectly by noting changes in the COD of test solutions.

Initially, the relationship between toluic acid concentration and COD values was established. Toluic acid concentrations between 18 and 60 ppm produced COD concentrations approximately 2.5 times as great. Thus, reductions in COD values could be directly correlated to reductions in toluic acid concentrations, assuming that only the acids contributed to the solution COD.

Appropriate aerobic and anoxic salts solutions (Skladany and modified from Shelton and Tiedje 1984, respectively)[4,5] were spiked with toluic acids (20 ppm of each isomer), and a biological inoculum consisting of a mixture of commercial organisms and a landfill soil suspension was added. Aerobic samples were continuously mixed at room temperature, while anoxic samples were flushed with nitrogen gas and incubated at room temperature with occasional mixing. Negative controls, desgiend to test for abiotic toluic acid loss, received mercuric chloride as a microbial poison. Samples were tested periodically for pH, inorganic nutrients (nitrogen and phosphorus), and COD.

During the 37-day aerobic study, the pH remained relatively unchanged at approximately 7. However, the pH in the anaerobic bottles rose from 7 to above 9.5 within 11 days from the start of the experiment. The high pH values fall outside of the 6 to 8 range normally thought to be optimal for biological growth. Throughout the aerobic and anoxic experiments, nitrogen (as ammonia) and phosphorus (as orthophosphate) concentrations remained at favorable levels.

However, an unforeseen problem arose with trying to indirectly quantitate toluic acid loss by measuring COD. First, the bacterial inoculum contributed a large amount of COD to each sample bottle. The toluic acids alone should have contributed only 150 ppm of the approximately 800 to 1000 ppm of COD present in each starting sample. Thus, it was impossible to state unequivocally if COD reduction was due to toluic acid biodegradation or the metabolism of other readily biodegradable materials. In addition, soluble COD samples exhibited large variations in concentrations. These results made it impossible to comment directly on toluic acid biodegradation from the tests. Results from the anoxic biodegradation experiment suffered from the same unexplained COD variability.

Since COD changes could not be used to determine toluic acid degradation, two other methods were used to show that bioremediation would indeed work. First, samples from some of the aerobic bottles were inoculated onto purified agar-mineral salts plates containing 60 ppm isomeric toluic acids as the sole carbon source. The plates were incubated at room temperature, and the resultant microbial colonies counted. Approximately 1,000,000 cells per ml of test solution were able to grow using toluic acids as their sole source of carbon and energy. This indicated that organisms were present in the aerobic test liquid that could indeed metabolize toluic acids. Second, samples were sent out for specific toluic acid analysis at the end of the project.

All aerobic samples showed nondetectable (less than 0.5 ppm) isomeric toluic acid levels. The anoxic samples, however, showed little reduction in toluic acid concentrations. Ortho-toluic acid concentrations were reduced from 20 ppm to an average of 17.5 ppm, and combined meta and para-toluic acid levels decreased from 40 ppm to an average of 38 ppm. The high pH in the anoxic samples may have created environmental conditions unfavorable to toluic acid biodegradation.

FULL-SCALE BIOLOGICAL TREATMENT SYSTEM

At the completion of the laboratory tests, the client evaluated the results obtained and presented the conclusions to the state regulatory agency. Based on the information submitted, the agency approved the use of a biological treatment system to remediate the contaminated leachate. The system was originally sized to treat flow rates of up to 5 gpm.

A process diagram for the full-scale treatment system is presented in Figure 2. A 3000-gal tank is initially used to provide hydraulic surge protection and pH adjustment. Carbon dioxide is used for pH control. If a pH excursion occurs, an automatic valve will stop all forward liquid flow until pH limits have been reestablished. A 5 gpm pump is used to provide mixing within the tank and to transfer liquid to the bioreactor. Additional piping is provided to permit the pH adjustment tank to overflow (if necessary) into the 20,000-gal chest located in the basement. This condition could occur with excess water flow or from maintenance requirements. A small electric pump is provided to recover the water from the chest.

After pH treatment the water is pumped through an automatically controlled flow valve into the biological reactor. The reactor (6 ft in diameter and 9 ft in height) is filled with a high-surface area inert plastic support media for the growth of bacteria. Air and water headers ensure that these materials are evenly distributed throughout the reactor. Oxygen (in the form of air) is supplied to the reactor to support aerobic metabolism. After treatment, the water flows by gravity from the reactor into a sump tank equipped with a bag filter and sump pump.

As part of the treatment process, biological solids (<0.5 lbs/day) are generated, which must be filtered to prevent plugging of downstream equipment. Periodic cleaning of the bag by dumping and washing was originally thought to be sufficient to keep solids from downstream operations. After the initial startup, a quiescent tank was added to the process stream in order to settle solids before they entered the bag filter. The solids generated are classified as a nonhazardous waste. Sludge blowdown from the reactor will also occasionally be required.

After filtering, the water is treated with activated carbon in the carbon

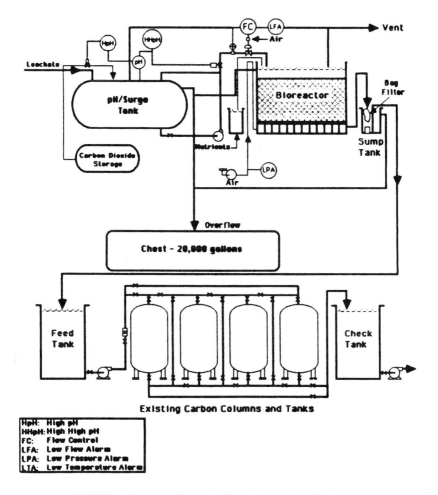

FIGURE 2. Process diagram for leachate treatment system.

columns used previously. Carbon polishing is used to remove any trace chlorinated organics not destroyed by biological treatment or by air stripping from the bioreactor. The effluent water is collected in an existing 3000-gal storage/check tank for analysis before final discharge.

SITE OPERATING CONDITIONS

Two factors had to be considered during the evaluation of the suitability of using biological treatment at this site: weather and operating personnel requirements. The site is located in the northeast United States, and the proposed treatment system had to be able to cope with low air temperatures during the winter months. In addition, the facility was closed, and a single caretaker was left to monitor the entire physical plant. It was hoped that this person could be used to operate the leachate treatment system.

The weather problem was overcome by placing the biotreatment system inside one of the unused buildings. In addition, a simple enclosure (equipped with a duct to the existing gas-fired heater) was constructed for the bioreactor and carbon columns. This heating system ensures that the temperature within the bioreactor remains above the 60°F necessary for bacterial growth. The unheated surge tank was also painted half black for solar gain.

The second major obstacle to overcome was the unavailability of a full-time operator to run the treatment system. Most people feel that expensive operator attention is one of the major drawbacks to biological treatment systems. At this site, the selection of a self-regulating submerged fixed-film bioreactor significantly lowered the amount of required operator attention. To reduce this requirement even further, an automatic monitoring system was installed via the existing ADT alarm system. High pH, low flow rate, low blower pressure, and low reactor temperatures are continuously monitored. The biological processes within the reactor are self-regulating, and require no human monitoring or adjustment. The only concern is about the ancillary equipment with moving parts, such as the pumps and blower. These items are all tied directly into the alarm system. With this setup, the caretaker not only knows when something is wrong at the plant, but he knows specifically which mechanical unit is not operating. Thus, the single site caretaker is able to add responsibility for the leachate treatment system to his other tasks.

FULL-SCALE SYSTEM RESULTS AND EXPENSES

The full-scale remediation system has been in operation for more than a year. Figure 3 summarizes the COD operating data for the first 270 days of operation (day 0 corresponds to July 14, 1987). As can be seen, the effluent COD concentrations have been very consistent, with the exception of one time period after six months of operation. The high COD effluent at that point was preceded first by a time of low influent COD and then by a spike of high influent COD. Fixed-film reactors are able to overcome moderate changes in influent organic concentrations without an adverse effect on effluent water quality. However, in this case the influent COD approximately doubled from 350 ppm to 700 ppm. Under these conditions the bacteria within the bioreactor needed some time to adjust to the increase in organic loading. Even during this time, all effluent discharge criteria were met.

Toluic acid analyses were run quarterly. Bioreactor influent and effluent

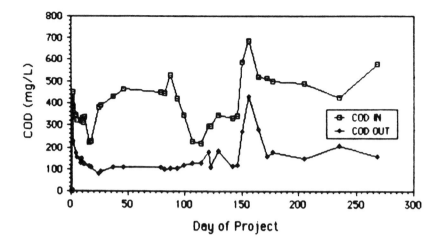

Day of Project

FIGURE 3. Bioreactor COD data.

Table 1. Bioreactor Toluic Acid Treatment Data (in ppm)

Contaminant	Sample Date		
	8/20/87	1/18/88	4/13/88
Influent ortho-toluic acid	43	79	71
Effluent ortho-toluic acid	<0.5	<0.5	0.05
Influent meta & para-toluic acids	1	25	45 (meta)
			3 (para)
Effluent meta & para-toluic acids	<0.5	<0.5	<0.078 (meta)
			<0.052 (para)

concentrations are summarized in Table 1. Influent ortho-toluic acid concentrations ranged from 43 to 79 ppm; effluent concentrations were less than 0.5 ppm, the normal detection limit for the compound. Combined meta and para-toluic acid concentrations ranged from 1 to 48 ppm. Again, effluent concentrations were less than 0.5 ppm. Total toluic acid destruction was therefore consistently greater than 98.5%. Thus, the nondetectable toluic acid concentrations obtained in the aerobic laboratory treatability study were confirmed in the full-scale biotreatment system.

Prior to choosing the biological treatment system, an economic analysis was performed for several treatment options. The off-site leachate disposal costs were $0.20 per gal, on-site activated carbon treatment costs were $0.08 per gal, and on-site biological treatment costs were estimated to be between $0.003 and $0.006 per gal. Biological treatment was clearly the most cost-effective means for treating the leachate under the given set of conditions.

Actual full-scale equipment costs totaled $37,580, including $21,800 for the bioreactor system and $13,900 for installation. Winterization of the bioreactor tank and carbon columns contributed $2000 toward the total. Daily operating expenses for the bioreactor are limited to approximately $5.40 for electricity to run the blower. The logical progression from problem analysis to laboratory testing and advanced process design resulted in the efficient and cost-effective biological treatment of toluic acid contaminated leachate.

ACKNOWLEDGMENTS

I would like to thank our industrial client for permission to use its project and information in this case history. I would also like to thank George J. Skladany for compiling this information and for his review of the article.

REFERENCES

1. **Alexander, M. and B. K. Lustigman.** "Effect of chemical structure on microbial degradation of substituted benzenes," *J. Agric. Food Chem.* 14: 410-413 (1966).
2. **Alsop, G. M., G. T. Waggy, and R. A. Conway.** "Bacterial growth inhibition tests," *J. Water Pollut. Control Fed.* 52 (10): 2452-2456 (1980).
3. **Evans, W. C.** Biochemistry of the bacterial catabolism of aromatic compounds in anaerobic environments," *Nature,* 270, 17-22 (1977).
4. **Shelton, D. R. and J. M. Tiedje.** General methods for determining anaerobic biodegradation potential. *Appl. Environ. Microbiol.* 47 (4): 850-857 (1984).
5. **Skladany, G. J.** Characterization of a mixed microbial culture able to degrade nitrilotriacetic acid and 2-chlorophenol," Master's Thesis, Clemson University, Clemson, SC (1984).

CHAPTER 9

Laboratory and Pilot Plant Evaluation of Ultraviolet (UV)-Oxidation Treatment Methods

Evan K. Nyer and Paul Bitter

INTRODUCTION

One of the most exciting technologies that is emerging as a viable treatment technique for ground water is UV-Oxidation. This is a technology that promises to destroy many organic compounds in a short amount of time and at a reasonable cost. EPA's Superfund Innovative Technology Evaluation (SITE) program has evaluated this technique and has published positive results.[1] Several companies have set forth on major marketing campaigns to promote UV-Oxidation to regulators, industries, and consultants in the groundwater field. The result has been that several UV-Oxidation projects are now proposed and some are being installed for treatment of ground water.

The problem is that, like most technologies that become "hot", too much is being promised and expected. We seem to take a technology with potential and then expect it to be able to work in every situation. The treatment systems

ISBN 0-87371-731-7
© 1992 by Lewis Publishers

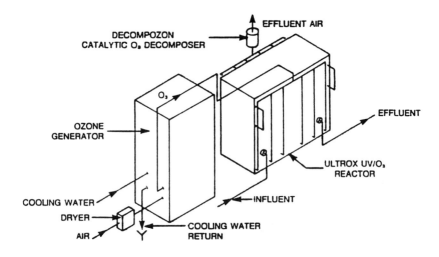

FIGURE 1. UV/Ozone Process Flow.[2]

are installed everywhere. The pendulum swings all the way back when some of the field units do not work. If this technology is to be used in the ground-water remediation field, we need to appreciate its capabilities and limitations. The purpose of this chapter is to help put this technology back into perspective and to provide some methods for more rounded evaluations of UV capabilities.

There are two basic forms in which UV-Oxidation is being applied: UV-Ozone and UV-Peroxide. Figure 1 shows a typical set up for UV-Ozone.[2] Figure 2 shows a typical set up for UV-Hydrogen peroxide.[3] Both systems use an oxygen-based oxidant, ozone for the first and hydrogen peroxide for the second. UV light is used in conjunction with the oxidant. The UV light bulbs are placed in the reactor where the oxidant comes into contact with the contaminants of the ground water.

While ozone and hydrogen peroxide are both strong oxidizing agents, their effectiveness increases dramatically when stimulated by UV light. Figure 3[2] is an example of the difference between oxidation with the oxidant alone and the oxidant stimulated with UV light. Similar types of increases are seen with UV and hydrogen peroxide. In both cases the key to fast reaction times is the UV light bulbs. However, the bulbs cannot come into direct contact with the water. They are normally covered by a quartz tube. The quartz protects the bulbs, but allows the UV light to enter the water unaffected.

The main difference between the two designs is the type of oxidant and the method of application. Ozone is an unstable gas. It must be added to the reactor as small bubbles and must be produced at the site. The UV-Ozone system includes an ozone generator. The ozone is sparged into the reaction tank below the UV lights. This creates a gas stream that must be evaluated for ozone and volatile organic compounds. The design in Figure 1 and the unit that was studied under the SITE program use "low" intensity UV bulbs.

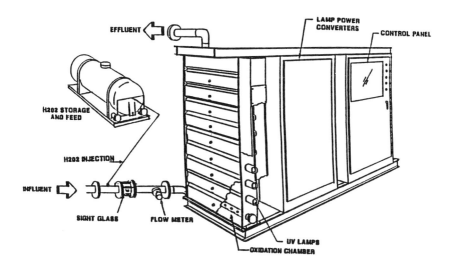

FIGURE 2. Equipment arrangement and process water flow for the UV/H202 System.[3]

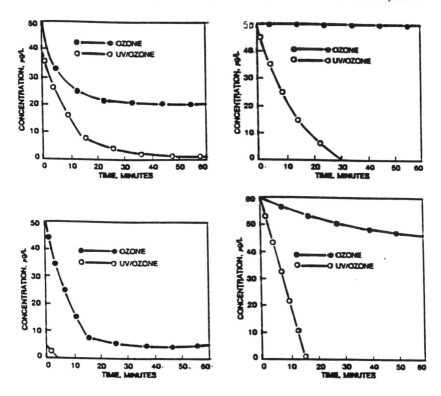

FIGURE 3. Effects of UV light on oxidation of organics.[2]

By comparison, hydrogen peroxide is a relatively stable liquid that can be delivered and stored on site. The peroxide is metered into the influent and the main reaction occurs within the reaction tank where the UV light is present. The design in Figure 2 uses "high" intensity UV bulbs. The reaction tank can be pressurized due to the lack of gas transfer requirements that would otherwise apply to ozone gas.

SITE PROGRAM

The two main objectives of the SITE demonstration test were to: (1) determine the operating conditions needed for the effluent to meet applicable permit limitations, and (2) to develop information required to estimate operating costs for the treatment system. My personal opinion is that the demonstration did not meet either objective. The study did present an interesting and useful comparison of the volatilization of some of the chlorinated ethanes (that proved to be a possible significant removal process), vs the chemical oxidation of the other volatile organic compounds that were tested.

The program was made up of 13 trial runs. Several parameters were studied: residence time, hydrogen peroxide dosage, ozone dosage, number of UV bulbs operating, and influent pH. These were then related to destruction and volatilization of the VOCs.

The demonstration was able to relate the effect of these parameters on the chemical reaction and the volatilization of the test compounds. But, all results were based upon short-term tests. Once the trial run reached three residence times, the data were collected and the run terminated. Therefore, the longest trial run was less than four hours. The only objective that this limited testing accomplished was to relate the reaction rates in a laboratory reactor to the reaction rates in the full-scale configuration.

One of the most important parameters as far as performance and costs was not considered during this trial — time. The demonstration took the results obtained in less than four hours and extrapolated them to over 8500 h (1 year). There is no guarantee that the reactor would sustain this performance over extended periods of time.

FULL-SCALE CASE STUDY

When we look at the data from an actual full-scale installation, we can obtain a better idea of the true performance and costs of this technology. The system used for this discussion involves a site where ground water is contaminated with organics: acetone, benzene, chlorobenzene, ethylbenzene, tetrahydrofuran, toluene, and xylene. The full-scale UV-Peroxide oxidation unit was designed for 15 to 20 gpm, 200 ppm of H_2O_2, and 180 kW UV power. The unit has been in operation for one year with an anticipated four additional years of operation. Data used for this review were taken over a 14-week period between December 1989 and March 1990.

The UV-Peroxide oxidation bench-scale study originally conducted for this site was successful in meeting the established treatment goals for all compounds, even for a worst case scenario. Based on the analytical data obtained from the bench-scale test, it was assumed that the full-scale UV-Peroxide unit would also be capable of treating the ground water to the treatment goals for all of the above contaminants. As shown in Table 1, this was not the case. The average concentration of benzene, for example, after UV treatment was 8,500 times greater than the concentration resulting from the bench-scale study and 340 times greater than the treatment goal. Even the lowest concentration detected for benzene, after the UV lamps were cleaned, exceeded the treatment goal of 5 ppb and the bench-scale results. The average concentration for two of the seven compounds exceeded their treatment goals. The percent removal efficiencies for the bench-scale and the full-scale units are included in Table 2.

The values listed for the full-scale unit were not only variable from week to week but also did not exceed any of the bench-scale unit removal efficiencies. The initial reasons behind the drop in performance was due to a lack of operator skill and the presence of iron in the ground water. As a result, a pretreatment system was designed and implemented to handle this problem. A pretreatment unit was installed by October 1989 to remove the constituents in order to improve system performance. The full-scale data previously discussed included the pretreatment unit.

Part of the problem with the operation of the UV-Hydrogen peroxide unit goes back to its design phase. When the full-scale unit was designed it was not known whether iron would occur at a concentration high enough to cause problems with the UV unit. The initial iron concentration was 15 ppm. The dissolved iron being oxidized by the hydrogen peroxide precipitated causing an orange coloration of the ground water and a coating on the quartz tubes. When the UV lamps are coated, the UV intensity decreases because it becomes more difficult for the radiation to penetrate to the water, thus reducing the system's efficiency. A pretreatment system was designed and implemented to remove the iron precipitation which created an additional waste stream.

A heavy coating was still occurring on the UV lamps even after the precipitated iron was removed. It was determined that this coating was a residual biological floc formation resulting from microorganisms in the ground water. This coating was not previously noticed because the biological floc initially was bound up within the iron floc in the oxidation chamber. In order to eliminate this problem, the primary oxidant was changed, but then a calcium carbonate scale formed on the UV lamps, still causing reduced efficiencies. An acid feed system was designed to combat this problem.

Even with these modifications, the performance data were not consistently good. Figure 4 is a graph of the average effluent concentration vs time. This figure shows that when the quartz tube was cleaned, the effluent concentration

Table 1. UV Treated Water Concentrations

Contaminant	Effluent Bench-Scale Results (ppb)	High Value (ppb)	Effluent Full-Scale Results[a] Average Value (ppb)	Low Value (ppb)	Treatment Goals (ppb)
Acetone	0.2	10,000	1,700	250	—[b]
Benzene	0.2	12,000	1,700	19	5
Chlorobenzene	1.1	5,200	730	11	100
Ethylbenzene	0.4	1,700	230	12	700
Tetrahydrofuran	0.2	6,300	1,200	50	1,400
Toluene	3.2	9,300	1,300	19	2,000
Xylenes (Total)	7.0	41,000	4,100	10	10,000

[a] Analytical data taken over a 14-week period between December 1989 and March 1990.
[b] No criteria have been established for acetone in drinking water since no health effects have been demonstrated at the concentrations thus far tested.

Table 2. Percent Removal Efficiencies

Contaminant	Bench-Scale Unit % Removal	Full-Scale Unit % Removal Week Number[a]											
		1	2	3	4	5	6	7	9	10	12	13	14
Acetone	99.99	98	96	0	95	0	93	94	95	95	97	96	98
Benzene	99.997	87	95	0	99.8	87	99	92	91	99	99	98	97
Chlorobenzene	99.995	87	95	10	99.8	90	99	93	91	93	99	98	97
Ethylbenzene	99.993	88	99	23	99	91	97	95	93	95	98	98	98
Tetrahydrofuran	99.997	73	99	17	99	68	96	87	72	74	96	90	89
Toluene	99.99	89	97	15	99.8	92	99	95	93	94	99	99	98
Xylenes (Total)	99.99	—[b]	95	94	99.96	93	99.6	94	97	93	99	99	97

[a] Analytical data taken over a 14-week period between December 1989 and March 1990. No data were taken for weeks 8 and 11 because the UV unit was not in operation.
[b] The concentration increased by 9%.

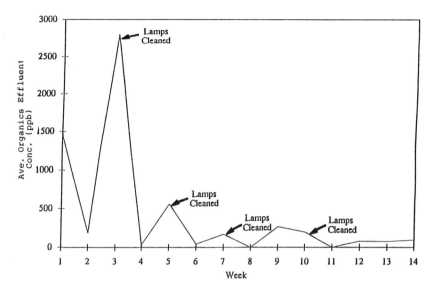

FIGURE 4. Average effluent organics from a full-scale operation. Note: No data was obtained for weeks 8 and 11 because the UV unit was not in operation. The UV lamps were cleaned between weeks 3 and 4, 5 and 6, 8 and 9, and 10 and 11, as indicated by the arrows.

decreased. Cleaning of the quartz tube was performed approximately twice per month and required the UV unit to be shut down for an eight-hour period.

A cost comparison (Table 3) between the initial estimate and the actual costs for the full-scale UV-Peroxide oxidation unit reflect a 24% increase in cost for additional and unanticipated modifications. The bulk of these costs was based on installing a pretreatment system, procuring additional chemicals, and requiring a greater amount of operator attention.

PROPER EVALUATION TECHNIQUES

What went wrong here? The proper tests were performed to show that the organics in the ground water would be rapidly destroyed by the UV-Peroxide system. The full-scale unit design had been previously tested in several field installations and found to be of optimum design.

The answer is that the wrong questions were asked when the system was originally designed. Previous full-scale unit testing only showed that the hydrogen peroxide, UV light and contaminated water would come into optimum contact. The laboratory work only showed that the pure chemistry of UV-Peroxide would work.

Long-term operation was completely ignored in all of this work. As with the SITE program, everyone assumed that if it works for four hours, it would work for the whole year. The full-scale installation showed that this was a bad assumption. The quick evaluation failed to predict total cost of treatment

Table 3. Cost Comparisons

	Initial Estimate ($)	Actual Costs ($)
Capital	275,000	275,000
Operation and maintenance (1st year)	12,000	72,000
Pretreatment	0	28,000
Chemical addition[a]	0	20,000
Granular activated carbon[b]	72,000	72,000
Operator attention[c]	22,000	33,000
	(14 hr/wk)	(21 hr/wk)
Total	380,000	500,000

[a] The cost includes labor and is on a per year basis.
[b] The cost includes carbon unit rental, carbon, and carbon regeneration and is on a per year basis.
[c] The operator's salary is $30/hour and is on a per year basis.

and, more importantly, the short-term tests failed to predict the performance of the full-scale system.

I have now worked with several UV-Oxidation systems, and the problems reported with this full-scale system are not unusual. The problem seems to be centered on the quartz tubes. Many chemicals and minerals in the ground water coat out on the quartz tubes and prevent the UV light from getting into the water.

This should not be unexpected with groundwater cleanups. As many of you know already, we normally do not use shallow aquifers for drinking water, because they are normally "dirty". Most groundwater contamination is in shallow aquifers. There will be many natural chemicals in the water along with the contaminant that will interfere with any groundwater treatment process. We have discussed these interferences in the context of treating ground water with air strippers many times in this book.

Does this mean that we should abandon this technology for groundwater cleanups? No! This is an important new technology that has great potential. We must change our method of evaluation of the UV-Oxidation system.

The laboratory methods that are being used are reasonable. We must remember that the results only reflect what the optimum reaction rate can be. The full-scale designs are good. However, with new companies bringing variations to the market, the designer will have to evaluate the field data from the various designs.

The main evaluation that needs to be added is the effect of the ground water on the quartz tubes. Right now, the only way to do that is to run a long-term (1 to 3 months) pilot test on the actual ground water to be treated. Hopefully, someone will develop a short-term test for the quartz tubes in the near future. But, for now, continue your evaluations of UV-Oxidation. Add a long-term pilot test to your program. This simple change will allow us to apply this important technology without installing units that give it a bad reputation.

REFERENCES

1. **Lewis, M. K., Topudurti, G. Welshens and R. Foster.** "A Field Demonstration of the UV/Oxidation Technology to Treat Ground Water Contaminated with VOCs," *Control Technology* — Air & Waste Management Assoc., 40 (4): 540-547 (1990).
2. **Bernardin, F. E., Jr.** "UV/Peroxidation Destroys Organics in Groundwater," 83rd annual meeting of the Air and Waste Management Association, Pittsburgh, PA, June 24-29, 1990.
3. **Fletcher, D. B.** "UV/ozone Process Treats Toxics," *Waterworld News,* Nov. 3, No. 3, May/June 1987.

SECTION VI

Selection of Treatment Alternatives

This section discusses combining all of the previous selections and selecting the treatment methods that will be used to remediate the site. The steps that we follow parallel the sections of this book:

1. Define treatment parameters.
2. Use the properties of the compound to perform a preliminary design.
3. Perform laboratory and pilot tests.
4. Select the unit operations.

We must start with valid data to define the treatment requirements. Detailed contaminant analysis must be combined with information about the natural components (i.e., iron, hardness, etc.) of the soil and aquifer.

The contaminant properties are then used to reduce the selection of potential treatment technologies. The contaminant properties helps us to select technologies that have the best chance of removing the contaminants economically from the soil and the aquifer.

The contaminant properties will also help to define what we do not know about contaminant treatment. The laboratory is the next step to answering these new questions. If the laboratory tests are designed correctly, we can derive the information about the contaminants and the technologies that will remove them from the site. Pilot plants can then be used to define further how these technologies will perform under field conditions.

Once we have chosen the technologies that we want to apply for the remediation, we have to evaluate how these technologies will be applied. Groundwater and soil remediations are unique and the applications of the technology must encompass the specific requirements for cleanup.

All of these steps lead to the selection of the treatment system that will be used to remediate the site. Usually, the complete treatment system will include more than one technology. Also, one other factor must be considered before a final selection can be made — money (capital investment and operation costs). The final way to compare different ways to treat the contaminants is how much each of the methods will cost initially and in the long term.

This section will review the process of selecting the final alternative for three different scenarios. Technical and economic concerns are evaluated in each selection. The practical point to be made in this section is to make sure that all of the costs are considered. Operations, manpower and maintenance are important factors for the total cost of a long-term remediation.

CHAPTER 10

Treatment of BTXE Compounds at Low Flow Rates

Evan K. Nyer

In this chapter I will focus on a treatment analysis query submitted by
Joseph P. Nestor, a hydrogeologist with the Southern Pump and Tank Co. in
Charlotte, North Carolina. One of the firm's underground storage tanks had
leaked gasoline. The basic contaminants in ground water are benzene, toluene,
xylenes, and ethylbenzene (BTXE). The inquiry stated:

> The site in question is located in central South Carolina on the upper coastal
> plain. The sediments at the site belong to the Tuscaloosa Formation, which is
> described by Pooser and Johnson (1961) as principally unconsolidated, cross-
> bedded kaoliniotic, arkosic sands, and conglomerate sands inter-bedded with
> lens of kaolin. Drilling logs tend to confirm this description.
>
> On the site is a convenience store that was completed less than two years
> ago. In July 1987, free-floating gasoline was discovered in observation wells

ISBN 0-87371-731-7

within the tank excavation. A review of inventory records indicated a shortfall of approximately 1100 gallons of unleaded and leaded gasoline. The discharges have been stopped.

At present we have been pumping 2000 gallons a day from wells at the site with quick recharge. I would expect that we could pump 4000 or perhaps 6000 gallons a day at present. I hope that the shallow aquifer will begin to dewater in the area of extraction and these yields will decrease.

Regulations prevent the discharge of wastes into the subsurface by means of wells. The regulatory definition of waste would include bacteria, nutrients, and treated effluent.

Data sheets were provided with the letter. The design flow rate was set at 2 gal per min (gpm). Following is a summary of expected influent and the required effluent discharge concentration levels for each contaminant. Concentrations are expressed as parts per million (ppm) or parts per billion (ppb).

Compound	Influent Concentration (ppm)	NPDES Discharge Req. (ppb)
Benzene	36.1	53
Toluene	20.5	175
Xylenes (total)	8.7	82
Ethylbenzene	1.5	320

The water can also be discharged directly to a publicly owned treatment works (POTW) if the total BTXE concentration is below 4.5 ppm. There are no regulations on air discharges for this site.

Although Nestor has provided a complete description of his groundwater problem, there are several questions that I would ask before starting to design a treatment system. These questions include:

1. What is the total organic content of the ground water? There will most likely be other organic compounds present in the water. Several types of treatment (carbon adsorption and biological treatment, for example) are affected by all of the organics present in the water, not just the compounds that are of interest to us. I would suggest that a Total Organic Carbon (TOC), a Chemical Oxygen Demand (COD), and a Biochemical Oxygen Demand (BOD) analysis be run on a representative sample of the ground water. I will assume that there are no other compounds to be treated in this ground water except for BTXE.

2. Is there any free product in the ground water when it is pumped to the surface? The first step in any groundwater treatment system is to remove the pure compounds. I will assume that there is no free gasoline in the ground water.

3. What is the iron (Fe) content of the ground water? The water was initially analyzed for heavy metals (which were not detected), but not for iron. Iron in the water can precipitate when oxidized, and may have an adverse effect on

air stripping and carbon adsorption treatment systems. I will assume that there is no iron present in this ground water.

The first step that I always take before the design of a treatment system is to list the physical properties of the individual contaminants present.

Compound	Molecular Weight	Specific Gravity	Solubility ppm (@ °C)
Benzene	78	0.88	1780 (20)
Toluene	92	0.87	515 (20)
Xylenes (total of 3)	106	0.88	540 (25)
Ethylbenzene	106	0.87	105 (20)

The specific gravity of water is equal to one under standard conditions. Compounds with a specific gravity of less than one will float on water, while compounds with specific gravities greater than one will sink. The BTXE compounds will each float on the ground water.

The BTXE influent concentrations should next be compared to the observed maximum solubility concentrations of these compounds in water. Each of the contaminants is present at below its solubility limit, and I would not expect to find any free-floating BTXE layer.

The next step in the design process is to find the treatability property for each of the contaminants. From experience I know that all of these compounds can be treated by either carbon adsorption, air stripping, or biological treatment, and that they are the most cost-effective treatment systems. One of the reasons I chose this case is because it provides a good opportunity to compare the advantages and disadvantages of each of these technologies under low flow rate conditions.

Compound	Adsorbability mg/g (@ ppb)	Henry's Constant Atm·M³W/M³A	Biodegrade
Benzene	80 (416)	240	Yes
Toluene	50 (317)	330	Yes
Xylenes (each isomer)	70 (500)	71	Yes
Ethylbenzene	18 (115)	350	Yes

The preceding values are all within ranges that indicate that the three technologies can be applied to this groundwater contamination. However, I cannot provide general cut-off points based upon these values that any individual can use to eliminate a specific treatment method. Those evaluations must be made on a case-by-case basis using all of the specific site information available. As a first step, treatability data from unfamiliar compounds can be compared to data from familiar compounds. With some caution, conservative extrapolations can be made regarding the treatability of unfamiliar compounds in a preliminary design evaluation.

The next step in the decision-making process is to evaluate the treatment methods based on their capital and operating costs. This is difficult to do in great detail within the constraints of this chapter. First, I want to stress that all of the costs presented here are estimates and should not be applied to other field projects. In order for the reader to gain some appreciation for the costs of different treatment systems, I have tried my best to get cost estimates representative of real-world expenses. However, designing and pricing efficient and cost-effective treatment systems depends upon the knowledge and skill of the design engineer. This is not a cookbook process. The reader should not use the cost estimates and treatment processes presented in this article for similar projects without performing a proper engineering evaluation.

I am now ready to discuss advantages and disadvantages of the three possible remediation technologies.

CARBON ADSORPTION

Carbon adsorption is a separation process whereby the contaminants to be removed from the water are attracted to and held at the surface of activated carbon granules. Carbon adsorption does not permanently destroy the contaminants present. Contaminated carbon must be disposed of properly in order to minimize future adverse environmental effects.

Frequently, the highest cost involved in using a carbon adsorption system is the disposal of used carbon and its replacement with new carbon. A "rule of thumb" that I use for estimating carbon capacity is that 5 to 20 lb of carbon are required to remove 1 lb of contaminant. The exact usage rate is controlled by the adsorptivity of the organic and the effluent concentration required. To provide a specific usage rate value, I have consulted Robert Byron of Tigg Corporation, Pittsburgh, PA. Byron uses a computer model based on theoretical carbon adsorption rates that has been calibrated with real-world data. The computer program requires the names of the contaminants, their concentrations, and an expected water flow rate. It then provides the appropriate carbon utilization rate.

Byron's model showed that 6 lb of carbon would be used per day to remove the 1.6 lb of BTXE compounds present. To ensure saturation, two carbon units would be used in series. The effluent levels acceptable for the POTW or the NPDES permit are both relatively high, and are easy to attain using carbon. For the activated carbon system design, I will assume that two carbon units are used in series and that the carbon does reach saturation.

The operating costs for the carbon system come mainly from carbon replacement and disposal. New carbon costs approximately $1.25/lb for small quantities. Disposal of the used carbon at a regeneration center would cost aprpoxiamtely $0.75/lb, and transportation from South Carolina to the regeneration site is expected to require another $0.75/lb. Sending the carbon to a landfill would reduce the disposal cost slightly, but my personal opinion (one of the perks of writing your own book) is that spent carbon should not be landfilled, primarily for liability reasons. The total system operating cost would be $16.50 per day (6 lb of carbon at $1.25 + $0.75 + $0.75/lb). This translates into a yearly operating cost of $6,023.

The capital costs are based upon the size of the carbon vessel(s) used. Most engineers would base the vessel size upon the residence time of the liquid in the carbon. A 15-min contact time is considered the minimum residence time. The required residence time is best determined in laboratory or pilot plant studies. My experience has been that practical considerations control the size of small-scale carbon systems. I do not want to change the carbon more often than four times per year. This would require a minimum of 540 lb of carbon per vessel. The smallest standard vessel that I could find with greater than 540 lb of carbon contains 800 lb of carbon. The total residence time for two 800 lb carbon units would be four hours. Two carbon units (800 lb each) would cost about $8,400.

AIR STRIPPING

Air stripping is a separation technology that takes advantage of the fact that certain chemicals are more soluble in air than in water. By bringing large volumes of air in contact with the contaminated water, a driving gradient from the water to the air can be created for each of the compounds. Air stripping is not a destruction technology. It merely moves contaminants from a liquid phase to an air phase.

The main expense of air stripping is the capital cost of equipment. The capital cost is a function of the diameter and height of the required tower. For design purposes, a "rule of thumb" says that a liquid loading rate of 20 gpm/ft^2 of cross-sectional area is fine. Based on a 2 gpm flow rate, the resulting stripper for this application would be 4.3 in in diameter. This is too small a diameter for an air stripper. A second rule of thumb is that the ratio of the packing size to the diameter of the tower-should be at least 1 to 6. Using 2-in packing material, this would require a 12-in diameter tower. The smallest standard tower available is usually 10 in in diameter, and would result in a liquid loading rate of 3.7 gpm/ft^2 of cross-sectional area. At this flow rate, the good news is that the tower would be very efficient at removing the contaminants; the bad news is that not all of the surfaces on the packing would receive a constant water flow.

The compound that must be removed to the greatest extent during treatment is called the controlling compound. The controlling compound for this design would be benzene, which would require 99.85% removal for acceptable NPDES discharge (as you will see, the design cost will be the same for the POTW discharge). The tower height required for removing this compound would be 20 to 25 ft.* The exact height or diameter required does not really matter in this case because the price for any air stripping tower less than 14 in in diameter and less than 25 ft in height would be about $11,000, including the blower. Smaller tower sizes would not save significant amounts of money.

The main operating cost for the air stripping system would come from the power requirements of the blower. The blower would be $^1/_3$ horsepower (hp). Once again, this unit is so small that the actual air to water ratio is not a large consideration in sizing the blower. A $^1/_3$-hp blower is about the smallest reliable blower available. Assuming a cost of $.10 per kilowatt hour (kwhr), the daily operating cost for the system would be about $.60, or $219 per year.

This figure does not include any money for maintenance of the packing. Iron and/or degradable organics in the water can cause scaling and/or bacterial growth on the packing. When this occurs, the packing will have to either be cleaned or replaced. I am not going to include the cost of packing maintenance here, but the reader should be aware of these possible additional costs associated with air stripping.

BIOLOGICAL TREATMENT

Biological treatment uses the action of microorganisms to metabolize the contaminants present. Under aerobic conditions, contaminants may be completely converted to carbon dioxide, water, and additional bacteria. Thus, biological treatment is considered to be a true destruction process by which the contaminants are permanently remediated.

There are many ways to apply biological processes to degrade organics in water. There is not space to review all of the types of bioreactors available, but I recommend that a submerged fixed-film biological treatment design be used for this small-flow application. These units are very simple to run and can be left unattended for long periods during operation. The design to be discussed will be the same for either discharge requirement (POTW or NPDES).

A rule of thumb for this process design is that 1000 ft^3 of media are required to remove 60 lb of organics per day. With influent and effluent structures, air distribution, free board, etc., the bioreactor would be about 4 ft in diameter and 7 ft in height. The system would also require a 2-hp blower

---

* My thanks to Kevin Sullivan for providing these numbers.

and a nutrient feed system. The nutrient feed system would consist of a metering pump and a drum of ammonia and phosphorous in a 5 to 1 ratio. The capital cost for this entire treatment system would be about $12,500. The operating costs come from the power for the blower and the nutrient requirements. A 2-hp blower will cost about $3.60/day to operate. The nutrients are required at a carbon:nitrogen:phosphorus ratio of 100:5:1. Based upon the 1.6 lb/day of organics to be removed, and an off-the-shelf nutrient product, the nutrient cost would be approximately $.50/day. The total operating cost would be $4.10/day, or $1,497/year. The unit would produce about 10 to 15 mg/L of suspended solids. These levels should meet all discharge requirements for either the POTW or NPDES option.

SUMMARY OF COSTS

The costs that can be generally calculated for all situations were presented earlier. The following information summarizes the capital and operating costs for the three treatment systems:

Treatment	Capital Cost	Operating Cost (per year)
Carbon adsorption	$ 8,400	$6,023
Air stripping	11,000	219
Biological treatment	12,500	1,497

There are several costs not included in these figures. I have already mentioned costs that are unique to each technology, such as packing maintenance for air stripping systems. There are other cost categories that will apply to all of the technologies but will be controlled by local circumstances. Local variables, such as materials costs and building permits, cannot be reviewed in this chapter. Cost of installation and manpower expenses for operation will vary greatly across the country, but must be considered when determining the full cost of the treatment system.

I have assumed very simple designs for all three technologies. Their installation costs should be about equal, and are expected to be less than $5,000. Manpower requirements will vary with the technology. All three systems will require sampling and simple maintenance for the moving parts. The carbon system will require extra manpower for changing the spent carbon, and the bioreactor will require a small amount of manpower to change the nutrient drum. The important point is that the design engineer has to consider all of the costs (capital, operating, and manpower) associated with each project.

Finally, the total cost for the duration of the project should be the basis of the design, not only the first-year costs. The concentration of contaminating organics will decrease with time as the project progresses. The design process must account for changing site conditions, as changes in contaminant concentration or flow rate may have a dramatic impact on system operating performance and costs. I refer to these considerations as Life-Cycle Design. For a small flow system, only the carbon adsorption system will be affected by Life-Cycle Design considerations. The air stripper and the bioreactor will continue to cost the same as the contaminant concentrations decrease. The carbon usage, however, will be proportional to the total amount of organics present in the influent. The design engineer should estimate the length of the project and the rate of decrease of the organics (if possible) to more accurately determine the total costs of the cleanup project.

MAJOR FACTORS THAT WOULD CHANGE THE DESIGN

The design requirements presented in this scenario show air stripping to be the most appropriate and lowest cost treatment system for this ground water. There are several factors that, if changed, could significantly affect the costs of any treatment system.

An air stripper only separates the organics from the water. The air stripper will be affected by regulations that cover air emissions. If some type of air emission control would have to be added to the treatment system, capital and operating costs would increase significantly. Carbon adsorption and biological treatment systems can each treat the ground water without significant air emissions.

Lower effluent concentration requirements would add cost to the biological treatment system. The bioreactor can treat BTXE compounds to the 30 to 40 ppb range. If the effluent concentrations need to be below these levels, a posttreatment system (such as carbon adsorption) would have to be added. Lower effluent levels would have a minor effect on the air stripper or the carbon adsorption systems.

The carbon adsorption and biological treatment system would be affected by much lower influent concentrations. For the carbon system, less carbon would be used, significantly lowering the total cost of the project. The biological unit would be eliminated from consideration if the total influent organic concentration became too low (less than 1 mg/L). Air stripping would not be affected by lower influent concentration.

CONCLUSION

The conclusion the reader should draw from this chapter is that every site remediation is case specific, and must be carefully and individually evaluated.

Frequently, several different remediation technologies may be appropriate for use at a given site. When this is the case the process engineer should review several technologies (or combination of technologies) that can provide both efficient and cost-effective (capital, operating, and manpower) remediation. While air stripping was the best and lowest cost method for remediating this site, it would be erroneous to conclude that air stripping should always be used to remediate gasoline contaminated ground water at service stations.

CHAPTER 11

Total Dissolved Solids in Ground Water

Evan K. Nyer and Gregory Rorech

I am going to use one of the projects that I have been working on as the basis for this chapter.

This site is unusual as the main contaminants are total dissolved solids (TDS). The primary purpose of the groundwater treatment system will be TDS removal. There are minor amounts of volatile organic compounds (VOCs), but we will not dwell on their treatment. In all other aspects, this is a typical groundwater cleanup. The TDS was released to the ground water and the original source of the contamination has been eliminated. The TDS compounds are not naturally occurring. We expect the TDS concentration to decrease over a five-year period. Therefore, we used life-cycle design techniques for the groundwater treatment system. So, while the technologies are unusual, the engineering design methods remain constant.

ISBN 0-87371-731-7

SITE DESCRIPTION

A hydrogeologic investigation was conducted at the site to evaluate conditions for the installation of monitoring wells to comply with the requirements of the Resource Conservation and Recovery Act (RCRA). As part of the evaluation, monitoring wells were installed and soil and water samples were collected and analyzed. The investigation revealed that most of the site is underlain by 20 to 30 ft of fine-grained surficial sands with isolated areas of clayey sand with low permeability. The surficial sands are underlain by a clay layer approximately 50 to 60 ft thick.

The groundwater quality investigation revealed that the surficial aquifer has been impacted by two unlined waste storage ponds situated by a waste water discharge ditch. The ditch and ponds received treated industrial waste water with permitted discharge levels of 2700 mg/L of total dissolved solids and 270 mg/L nitrate. Recent regulations mandated the remediation of ground water impacted from the permitted discharges.

DESCRIPTION OF COMPOUNDS IN GROUND WATER

TDS concentrations in the ground water were detected in the 10,000 to 20,000 mg/L range. The TDS consisted primarily of sulfates, nitrates, and sodium. In addition to the total dissolved solids, the surficial aquifer was also impacted by VOCs. However, remedial technologies for VOCs have been extensively evaluated, so we will limit our detailed discussion to remedial options evaluated for the TDS. A full study, including literature search, laboratory tests, and field pilot testing, was conducted on VOC removal. Air stripping was selected as the treatment method. The air stripper will be designed for a possible future vapor phase, carbon unit.

As part of the RCRA facility permit, corrective actions for ground water contamination are govered by federal and state primary and secondary drinking water standards.

The compounds controlling remediation of the site will be the nitrate and sulfate levels. Nitrate concentrations were detected from below 10 mg/L to almost 2000 mg/L, while the sulfate concentrations were in the 3000 to 5000 mg/L range. As dictated by RCRA facilities permit, groundwater remediation must continue until the nitrate concentrations are below 10 mg/L and the sulfate concentrations are below 250 mg/L. As with most surficial aquifers, iron is present and will require pretreatment.

DESCRIPTION OF PROCESSES EVALUATED
FOR REMEDIATION OF TDS

Various technologies were considered for remediation of the ground water at the site. Due to the ''mixed bag'' of contaminants and the high TDS concentration, electrodialysis and ion exchange were eliminated during the preliminary evaluation stage. Also, since plant operations personnel are familiar with the operation of evaporators and reverse-osmosis (RO) equipment, our treatment evaluation concentrated on those two possibilities.

Evaporators use a heat source to concentrate a solution or to recover dissolved solids by boiling out the solvent, which is generally water. Special provisions for separating the liquid and vapor phases and for removal of the precipitated or crystallized solids are provided. The different types of evaporators use various methods to accomplish these tasks.

The facility currently operates falling film and vapor recompression evaporators for treatment of their various waste streams. Due to scaling and the desire for a more economical evaporation system, the facility had previously conducted pilot studies on their various waste streams with a multieffect, multistage flash evaporator. Since the pilot studies showed favorable results for treating the various plant process waste streams, which are similar to the contaminated ground water, our evaluation concentrated on this type of evaporator.

RO separates a solute from a solution by applying pressure to force the solvent through the membrane. The selection of the membrane material, configuration, and operating conditions are critical to obtaining the desired results. The most common membrane materials are cellulose acetate, polyamide, and thin film composite. However, other materials do exist and research is currently being performed to develop superior materials. The various configurations for the membranes consist of spiral-wound, hollow fiber, tubular, and plate and frame. Specified operating conditions consist of pressure — which is generally dictated by membrane material — recovery rate — which is the percentage of feed water converted to product water — and flux rate — which is the flow rate of water that passes through a unit area of membrane. Flux rate of water through a membrane is proportional to the pressure differential across the membrane. The higher the pressure, the higher the flux rate for a given membrane. The flux rate also depends on the material thickness of the membrane and temperature of the feed water. Flux rates should be specified conservatively to provide for long-term operation of the membrane.

One of the advantages of the evaporator is that it would be a one-step process. The unit would produce clean water and the TDS would be in solid form. On the other hand, the RO system will produce clean water and a concentrated brine. The brine will require further treatment in order to be placed in a form that can be sent to final disposal.

GROUNDWATER TREATMENT SYSTEM

Because of the low permeability of the soils, more than 75 wells were required to completely capture the contaminated plume. From our investigation, we also realized that not all of the ground water would clean up at the same rate. The TDS concentration in the ground water recovered from some wells would drop below the level required for treatment sooner than ground water from other wells. The VOCs will also not be removed from the site at the same rate as the TDS. We expect to be required to continue treatment for VOCs after having progressed to acceptable TDS levels.

As can be seen by the following economic evaluation, any treatment method for TDS is expensive. This is especially true when compared to air strippers.

Life-cycle design methods allow us to optimize the use of low-cost treatment methods. For this site, there will be two major time variables. First, we expect the TDS plume to clean up in five years, while the VOC plume will take 10 to 15 years. Second, we expect that the number of wells requiring TDS removal will reduce over that five-year period.

The recovery system designed and installed at this site includes a complex valving and routing system so that the recovered ground water can be separated into low TDS and high TDS waste streams. This way, the low TDS waste stream would only require treatment for VOCs and not the more expensive treatment to remove TDS. The potential treatment paths are indicated in Figure 1.

COST ANALYSIS OF SYSTEM

Direct evaporation and RO were selected for cost evaluation. The RO system will be supplemented by evaporation for final solids disposal. It was assumed that combined flow from the ground water recovery wells would be 70 gallons per minute (gpm). Effluents from the evaporator would include condensate and solids. Treatment with a RO unit would only concentrate the dissolved constituents in the recovered ground water. The permeate from the RO would require further treatment to remove the VOCs prior to discharge. The concentrate from the RO unit would be discharged to a 10,000 gallons per day (gpd) evaporator.

Water quality data indicates that the design of the RO system will be controlled by the high nitrate concentrations and a dual-stage RO unit will probably be required. Since the nitrate concentrations are extremely high in two areas, we decided to evaluate a third process option for this design. It was assumed that the water from the high nitrate wells will be discharged directly to the evaporator while the other pump effluents would be treated with a single-stage RO unit.

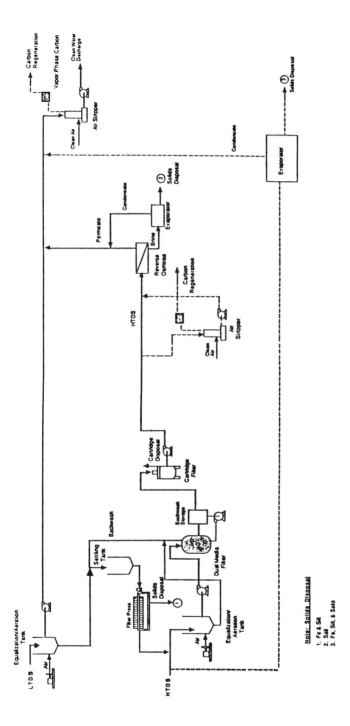

FIGURE 1. Process design for TDS treatment system.

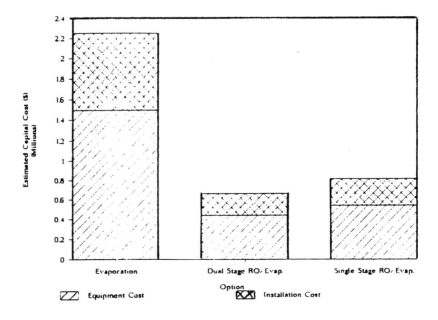

FIGURE 2. Demineralization capital cost.

Therefore, the three treatment options are:

1. Direct evaporator (two 50,000 gpd units)
2. Dual stage RO/evaporator (10,000 gpd unit)
3. Single-stage RO/evaporator (35,000 gpd unit)

Figure 1 shows the process design options including the pretreatment for the RO and the VOC treatment. Cost estimates for the demineralization systems were prepared for comparison purposes. The cost estimates are feasibility level and were used for guidance in project evaluation and implementation. Actual final project cost will depend upon labor, materials, and site conditions at the time of construction, as well as other market variables. Capital cost estimates were based on the assumption that the required steam for the demineralization systems is available on-site. The capital and operating cost estimates for the RO units include pretreatment equipment. The capital and installed cost estimates presented in Figure 2 indicate that the units are predominantly factory-assembled. Installed costs include the necessary site work for an operational system, taking site-specific conditions into consideration. As you can see, the RO/evaporator option is significantly less expensive than the direct evaporator.

Table 1 presents the estimated operating and maintenance (O&M) costs for the three remediation options. It was assumed that steam is available and that the existing power distribution system is sufficient with a usage rate of

Table 1. Demineralization O&M Costs[a]

Year	Direct Evaporation		Dual-Stage RO/Evaporation		Single-Stage RO/Evaporation	
	($/yr)	Cumulative ($)	($/yr)	Cumulative ($)	($/yr)	Cumulative ($)
1	429,000	429,000	267,000	267,000	338,000	338,000
2	384,000	813,000	254,000	521,000	322,000	660,000
3	384,000	1,197,000	254,000	775,000	322,000	982,000
4	307,000	1,504,000	254,000	1,029,000	322,000	1,304,000
5	307,000	1,811,000	254,000	1,283,000	322,000	1,626,000
Total O&M Cost		$1,811,000		$1,283,000		$1,626,000

[a] Based on 1989 dollars.

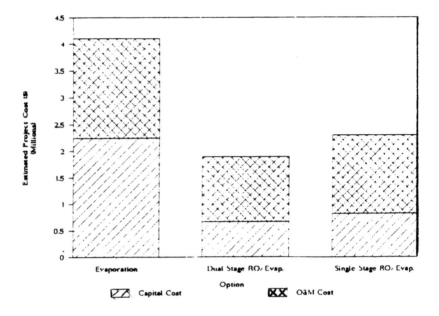

FIGURE 3. Demineralization project cost.

$0.08 per kilowatt-hour (kWh). In addition, it was also assumed that solids disposal costs would be approximately the same for the three remedial options. For the direct evaporation option, costs reflect the removal of one evaporator after two years of remediation. This evaporator would be used at the facility for treating other process waste streams.

Figure 3 presents the total project cost including capital, operating, and maintenance costs for the expected five-year TDS remediation project. The dual-stage RO/evaporator is the lowest cost system. It is half the cost of the direct evaporator and less expensive than the single-stage RO/evaporator.

The dual-stage RO/evaporator was selected as the preferred treatment system. The next step in the project is to run a pilot plant study on the RO system. This test will confirm the design parameters, and evaluate the performance of the various membranes on the specific compounds detected in the ground water.

CHAPTER 12

Treatment of Herbicides in Ground Water

Evan K. Nyer

In Chapter 10 we were concerned with the treatment of water contaminated with benzene, toluene, xylene, and ethylbenzene (BTXE). In this chapter we will discuss another interesting and relevant groundwater cleanup.

I have selected a case submitted by Keith B. Rapp, a hydrogeologist with Delta Environmental Consultants, Inc. in St. Paul, MN. He is working on an aquifer that has been contaminated with several herbicides. The data and information required for the site is summarized as follows:

Contaminants to be removed include — alachlor, atrazine, butylate, and metolachlor.

Flow to aboveground treatment system — A single pump-out well has an estimated volume of 25 to 50 gpm. The surficial aquifer in question is capable of producing significantly more water with minimal drawdown (hydraulic conductivity-ft/day).

Concentration of contaminants — The following information is supplied from a single well with the highest concentration of each contaminant:

ISBN 0-87371-731-7
© 1992 by Lewis Publishers

• alachlor	1800	μg/L (ppb)
• atrazine	3000	μg/L (ppb)
• butylate	3.5	μg/L (ppb)
• metolachlor	1300	μg/L (ppb)

The quantity of each contaminant spilled is unknown; however, it is estimated that the mass of each compound is less than 2 lb.

Effluent requirements — Currently, the state of Wisconsin does not have enforcement standards for these herbicides. The Wisconsin Department of Natural Resource (WDNR), however, has proposed the following criteria for these compounds:

• alachlor	0.15	μg/L (ppb)
• atrazine	0.35	μg/L (ppb)
• butylate	200.00	μg/L (ppb)
• metolachlor	15.00	μg/L (ppb)

Regulations controlling the installation — The WDNR is fairly liberal in terms of treatment technology. An alternative water supply has been implemented for the region in question and the well is no longer a residuental water supply. Therefore, an in situ treatment technique is feasible. Some of the treatment techniques considered included granular-activated carbon filtration and sunlight breakdown of the herbicides. The aquifer under consideration lies in west-central Wisconsin, and is composed of a very coarse-grained sand. The surficial aquifer is unconfined, has a saturated zone of approximately 65 ft, and is bounded by a relatively impermeable unit at its base.

While I am familiar with the general removal and destruction mechanisms for pesticides and herbicides, I have never worked with these specific compounds before. Therefore, I started the background research for this article by contacting the companies that manufacture these products. Monsanto makes alachlor under the tradename of Lasso®, Ciba-Geigy makes atrazine and butylate, and ICI America produces metolachlor. I also obtained the August 1987 U.S. EPA technical report "Health Advisories for 50 Pesticides" (document PB88-113543, available from the National Technical Information Service, Springfield, Virginia). From these sources I was able to obtain the physical properties of the individual contaminants present.

Compound	Molecular Weight	Specific Gravity	Solubility in ppm (@°C)
Alachlor	270	1.13	242 (25)
Atrazine	216	1.19	70 (22)
Butylate	217	0.94	45 (20)
Metolachlor	283	—	530 (20)

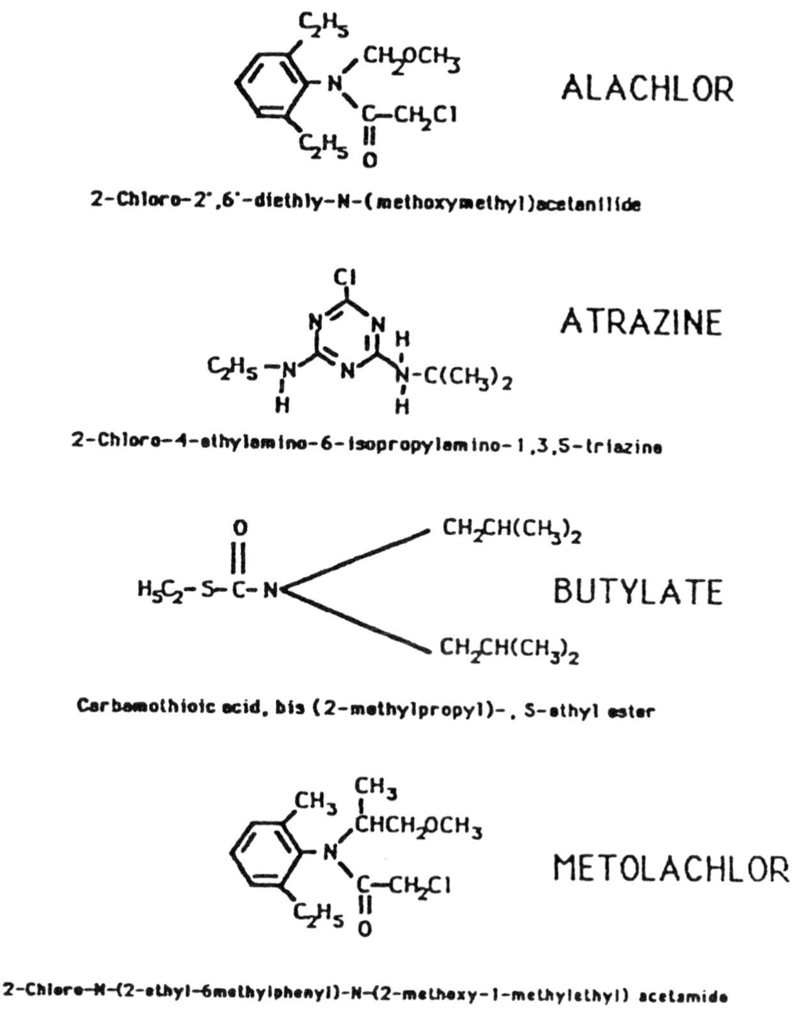

ALACHLOR

2-Chloro-2',6'-diethly-N-(methoxymethyl)acetanilide

ATRAZINE

2-Chloro-4-ethylemino-6-isopropylemino-1,3,5-triazine

BUTYLATE

Carbamothioic acid, bis (2-methylpropyl)-, S-ethyl ester

METOLACHLOR

2-Chloro-N-(2-ethyl-6methylphenyl)-N-(2-methoxy-1-methylethyl) acetamide

FIGURE 1. Structural formulas.

I also obtained the structural formulas for each of the compounds. I find the structural formulas to be helpful when I am discussing unfamiliar compounds with experts in different treatment technologies. The structural formulas for the four compounds are presented in Figure 1.

Finally, I asked representatives from each company for any available research or field data covering the removal of their products from water. I specifically discussed in situ treatment, carbon adsorption, and sunlight breakdown (photo decomposition), as mentioned by Rapp. None of the companies had any published work available on the treatment of their herbicides. Monsanto has ongoing research in this area. Two of the companies felt that in situ treatment and photo decomposition would not be able to clean the aquifer to the required cleanup levels. The representatives from each of the three companies felt that carbon adsorption would be the best treatment technology for remediating contaminated ground water.

From my experience, the only compound that I would expect to be readily biodegradable in an in situ cleanup would be butylate. Since butylate is already present at concentrations below the cleanup standard, in situ remediation of the aquifer would not be my first choice for a remediation strategy. However, if faced with a particularly large volume of soil or water to remediate, I would recommend that laboratory treatability studies be performed in order to more accurately test the action of microorganisms against the herbicides. If the laboratory results were encouraging, field testing could then be implemented.

While none of the manufacturers felt that photooxidation with visible light would be successful, the combination of ultraviolet light (UV) with an oxidizing agent (such as hydrogen peroxide or ozone) has been shown to effectively destroy complicated compounds such as herbicides and pesticides. Therefore, I have selected three different treatment methods on which to perform cost estimates. The three methods are carbon adsorption, UV light with hydrogen peroxide, and UV light with ozone.

Even though Rapp has provided a complete description of his groundwater problem, there are still a few questions that I would ask before starting to design a final treatment system. These questions include:

1. What is the total organic content of the water? All three treatment methods under consideration will be affected by the other organic compounds present in the water. Activated carbon will use some of its capacity to remove these additional compounds and will, thus, have less capacity available for the herbicides. The two oxidation methods will require an increase in oxidizing agent if other organic material is present. I would suggest that a total organic carbon (TOC) or a chemical oxygen demand (COD) analysis be run on a representative sample of the ground water. In this chapter I will assume that the only organics in the ground water are the herbicides.

2. What is the iron (Fe) content of the ground water? Iron can precipitate and clog the carbon bed of an activated carbon system. In addition, iron will also precipitate out onto the quartz tubes that are used with either UV-oxidation method. This precipitation will prevent the ultraviolet light from the bulbs from entering the water and catalyzing the desired reactions. I will assume that iron is not present in the water.

3. What is the total amount of material present in the aquifer? Rapp states that "less than 2 lb" of each compound is present. The concentrations from the single well pumped at 50 gpm would therefore clean the aquifer in less than one week. I suggest that he test the ground water during a pumping test to establish accurate concentrations of the herbicides. I will use the concentrations found in the one well at 50 gpm as my design basis.

I am now ready to discuss the advantages and disadvantages of each of the three possible remediation technologies.

CARBON ADSORPTION

Adsorption is a natural process in which molecules of a liquid or gas are attracted to and then held at the surface of a solid. Adsorption onto activated carbon is a physical process whereby the attraction is caused by the surface tension of the carbon. Activated carbon has up to 1400 m^2 of surface area per gram of carbon available for adsorption of organic molecules.

Carbon adsorption is a separation technique. The organic compounds present are removed from the water and are transferred to the surface of the carbon. The molecules are not changed or destroyed. Contaminated carbon must be properly disposed of or regenerated. For example, the compounds can be removed from the carbon and concentrated, destroyed at high temperatures under anoxic conditions, or the carbon can be buried in a hazardous waste landfill with the compounds still adsorbed. So, while the compounds are removed from the water, they have not been detoxified or destroyed. A complete carbon adsorption treatment system design must account for the final disposition of the organic compounds.

For the specific design of a carbon adsorption system in this case I contacted Mark Stenzel from the Calgon Carbon Corporation (Pittsburgh, PA). He informed me that Calgon had a research program on very similar compounds in progress. From that work, and other field installations, he was confident that carbon would be able to remove the herbicides to the required concentration levels.

Calgon uses a computer model to design carbon adsorption systems. The model recommended a carbon adsorption system consisting of two carbon units connected in series. Each unit would contain 2000 lb of activated carbon and, overall, the system would provide an average water-carbon contact time of 22 min. The computer design estimates that 6.7 lb of carbon will be required to remove 1 lb of total herbicides. (This is within my "rule of thumb" for carbon usage that states that 5 to 20 lb of carbon are needed per pound of contaminant removed.) At the anticipated design flow rate of 50 gpm, 3.7 lb of herbicide will be removed per day. Therefore, the carbon utilization rate will be approximately 25 lb per day.

The capital cost of this system would be $45,000. The operating cost would come mainly from the replacement of spent carbon. Calgon will remove the spent carbon and replace it with regenerated carbon for $1.25/lb plus transportation. I will assume that the transportation costs are the same as what I used in Chapter 10, approximately $0.75 per pound. This would result in a cost of $50 per day, and a yearly cost of $18,250. Stenzel also stated that they could provide a "temporary" treatment system of the same size for $100 per day. As the carbon within the temporary system became saturated, the entire 2000 lb unit could be sent back to Calgon and replaced. I will only use the permanent system to compare costs between the three treatment system designs in the cost summary section. However, the reader should be aware that potentially low-cost temporary systems may be available for short-term projects.

Comparing these costs to the carbon costs that were cited earlier, the capital costs are relatively higher and the operating costs are relatively lower. Calgon has a major advantage in that it takes back the spent carbon and reprocesses it. However, it is not my purpose to select the best carbon company in the United States. I want to expose you to the alternatives that exist for groundwater treatment and, also, to briefly explain the methods used to evaluate those alternatives. Therefore, I will not try to reconcile the different carbon costs, but simply use one firm's numbers for each case. I will continue to use different experts and companies in these evaluations so that you are exposed to the greatest variety possible.

OXIDATION WITH ULTRAVIOLET LIGHT AND HYDROGEN PEROXIDE

Oxidation technologies can be used to completely destroy the organics present in water. With the UV/peroxide system, high-intensity ultraviolet lights are used to catalyze the formation of hydroxyl radicals from hydrogen peroxide. The hydroxyl radical is the most powerful chemical oxidant available after fluorine. Under controlled conditions, the hydroxyl radical reacts with the contaminants present, oxidizing the chemicals to carbon dioxide (in the case of hydrocarbon contaminants) and halides (in the case of halogenated materials). There are no air emissions or waste by-products from the process if the reactions are carried to completion.

Unit selection is a function of the flow volume and time required in the oxidation chamber. The treatment process is dependent upon several variables, including the type and concentration of contaminants present, the waste stream flow rate, temperature, the peroxide and UV doses used, water pH, mixing, and catalysts present.

The specific design for a UV/peroxide system is very dependent on the company used to build the unit. There are no standard designs for this type of system. Therefore, I will not be able to make rule-of-thumb design calculations for this process.

I contacted Ronald Peterson of Peroxidation Systems, Inc. (Tucson, AZ) for help with the specific design for this groundwater problem. Peroxidation Systems uses a small number of high intensity UV bulbs for their designs.

They have had extensive experience in the oxidation of organics with their system both in the laboratory and in several field installations. They felt that there would be no problem in obtaining the required effluent limits. Peterson recommended the use of a contact tank measuring 2 ft wide, 8 ft long, and 5 ft high. The residence time within the reactor would be about 8 min. The power requirements for the system would be 30 kilowatts (kw), and the system would run 24 h/d.

The capital cost for this system would be $41,000. The operating costs would come mainly from the power for the UV bulbs and the hydrogen peroxide. At $0.06 per kw/h, I calculate the daily power cost to be $43. The peroxide costs are estimated to be $200 per month. Together, these expenses would come to a total of $18,000.

The operating costs were lower than I expected for this system. From previous work I would have guessed that the power requirements would have been two to three times higher. Peterson stated that a 10-gal sample was all that was needed for a laboratory treatability study to confirm the power requirements for oxidizing the herbicides. If there is time I would also recommend that a pilot plant be run to ensure that nothing in the actual ground water interferes with the transmission of the UV light.

OXIDATION WITH ULTRAVIOLET LIGHT AND OZONE

This remediation technology utilizes the strong oxidizing properties of UV light and ozone, a combination with destructive results superior to the use of either oxidant alone. The technology is suitable to influent organic concentrations in the ppm to low ppb range, and preferentially oxidizes halogenated hydrocarbons.

This technology is also a destruction process, eliminating future liability problems with the contaminants. The chemical reactions take place in side a reactor containing many UV bulbs. Water to be treated is passed through the reactor and past the bulbs. At the same time, ozone gas (generated by a separate ozonator) is bubbled into the water. The UV light and ozone act synergistically to oxidize the contaminants present.

Once again the specific design of this technology is dependent on the company that is supplying the equipment. There are no standard designs. I contacted a company experienced in this technique for help with the design. It has been studying the oxidation potential of ultraviolet light with ozone for more than 15 years. From the company's experience, it felt that there would be no problem removing the herbicides down to the required levels.

The company uses standard intensity UV bulbs held within a stainless steel tank. The ozone is bubbled into the bottom of the tank, and a header system evenly distributes the gas bubbles across the bottom of the reactor. It recommended that a 1500-gal contact tank, providing a 30-min residence time, be used for this application. The capital cost for the contractor and the ozonator would be approximately $180,000. The operating costs for the system would come mainly from the power requirements of the UV bulbs and the ozonator, and also the replacement cost of the UV bulbs (bulbs are typically replaced once per operating year). The company estimated that the operating costs would total $56 per day, assuming a cost for electricity of $0.06 per kw/h. This would require a yearly total operating cost of $20,440.

SUMMARY OF COSTS

The following summarizes the capital and operating costs for the three potential treatment systems:

Treatment	Capital Cost ($)	Operating Cost (per year) ($)
Carbon adsorption	45,000	18,250
UV/peroxide	41,000	18,100
UV/ozone	180,000	20,440

As before, these costs do not include permitting, engineering, transportation, installation, manpower, and analytical requirements. All of these costs relate to local conditions and requirements, and cannot be accurately calculated for use in this article. However, I would expect that the installation costs for the two oxidation systems (UV/peroxide and UV/ozone) would be slightly higher than the installation costs for the carbon system.

Based upon the capital and operating costs presented in this article, I would have a hard time choosing between the carbon adsorption system and the UV/peroxide system. I would recommend that laboratory and/or pilot plant studies be conducted with one (or both) systems prior to final treatment system selection.

Lastly, the expected duration of the project must be considered before the final decision is made. First, the contaminants present, their concentrations, and the groundwater flow rate may change over the life of the project. Any remediation system selected must be able to successfully adapt to these changing site conditions. I call this engineering flexibility "life-cycle design". Second, the total cost of the remediation effort is the sum of both the capital and operating expenses incurred throughout the life of the project. This expense is frequently much different (and greater) than the estimated first-year costs. Life-cycle design and total project expenses will tend to make activated carbon treatment systems preferable for use on short-term projects, while long-term projects may find the UV/peroxide process to be the more cost-effective treatment system.

I want to stress that all of the costs presented here are estimates and should not be applied to other field projects. For the reader to gain some appreciation of the costs of different treatment systems, I have tried to get cost estimates to represent real-world expenses. I have to assume that the information received from all parties is true and representative. The reader should not use the cost estimates and treatment processes presented here for similar projects without performing a proper engineering evaluation.

ACKNOWLEDGMENTS

I would like to thank George Skladany for his assistance in gathering the data used in this chapter and for his review of the final document. I would like to thank the people already mentioned in the text, as well as Bill Frantz, for their data.

SECTION VII

Practical Problems

I stated in the introduction of this book that one of the main reasons that all of these articles were put together in one place was to provide the reader with a concentrated reference for practical problems encountered with groundwater and soil remediation. While all of the chapters include practical as well as technical discussions, this entire section is devoted to practical problems.

When I started to contribute to *Groundwater Monitoring Review* in 1987 there were very few operating groundwater treatment systems. And, those that were operating had not been running very long. I felt that it was important to provide the type of practical information that comes from experience. I installed my first groundwater treatment system in 1983, and had installed over 30 systems by the time I started to write the articles. Therefore, I had first-hand knowledge on the unexpected problems that can be encountered when actually installing these systems.

I quickly realized that one of the main problems that would be encountered in this new field would be the lack of experience. Anyone could look up in a textbook or technical article how a specific technology could be applied to the removal of a contaminant from water or soil. However, there were no designers with sufficient experience to review the designs to prevent repeated mistakes. That was the main reason I started to concentrate on the practical nature of remediation designs. The surprise that I have had years later is that we are still not considering the practical parts of design. Few authors write about this area, and we have been expanding so fast that we still do not have enough experienced technical personnel to review all of the designs being created. This is especially true when we review designs done by the ''low cost bidder''.

The message of this section is to the managers that will be stuck with these treatment systems long after the investigation and remediation design teams have left the site. The manager must make sure that while the treatment system is being designed, the mundane, practical factors are considered. These factors are the ones that will make the treatment system technically fail or cost more than originally planned.

CHAPTER 13

Some Practical Problems

Evan K. Nyer

In this chapter I will provide some examples of practical problems that I have run across while applying groundwater treatment systems to the field. While I would call most of these problems unsophisticated, they can ruin a treatment system as easily as a major design flaw can.

SAMPLING

Everything that a design engineer does is only as good as the data that he receives. That data is based upon the samples and the analyses available for that ground water. There are several manuals that provide detailed information on the correct methods for sampling and analysis.[1-3]

ISBN 0-87371-731-7
© 1992 by Lewis Publishers

Logic

There is also the practical side of sampling. I do not want to tell you how many times I have received data in which the temperature of the ground water was measured in a laboratory (three or four this year alone). As you know, the temperature in the laboratory will not represent the true temperature of the ground water. However, if the person responsible for taking the sample does not understand the significance of the data developed, then he/she can make a simple mistake such as this.

On a practical level, all measurements that can be taken at the well should be taken. Temperature, pH, and dissolved oxygen can be measured in the well itself without bringing a sample to the surface. The person taking the sample should understand what is being measured. A small amount of time spent explaining will pay big dividends in accuracy.

Technique

The same holds true for how the sample is treated on its way for analysis. For example, the chlorinated compounds that are found in drinking water (TCE, PCE, vinyl chloride, etc.) do not stay in water for a long period of time. Turbulence while taking the sample or an air space in the sample container will cause a significant loss of the chlorinated compounds from the water.

I know that some of you are asking why are we so worried about low concentrations of these compounds if they do not stay in the water. One answer is: Good question. But, of course, the proper answer is that the public is concerned, and government agencies are responding to that concern. Whatever the reason, if we are going to treat the water, we need accurate information.

Static Versus Dynamic

One final area in sampling to discuss is what water from the well represents the "true" ground water. Currently, standard practice is to remove a certain volume of water from the well before the sample is taken. The stagnant well water is in equilibrium with the atmosphere, not the ground water.

I believe that flushing the well is the best method for getting a representative sample of the ground water at the location of the well. However, I have found that this method does not provide accurate data for a treatment process. The treatment equipment will receive water that is being pumped at a high rate from the well. The concentrations of the contaminants are usually lower in the pumped water. This makes sense. Once the pump is turned on, the water in the well is a combination of the ground water in the entire zone of influence. If the pumping rate is designed to capture the entire plume, then some of the water entering the well will be clean.

I recommend that samples be taken during the pumping tests. The data from these samples combined with the static well tests will provide a more accurate design basis for the treatment system.

This is different from what I call "life-cycle design". Over the life of the project, the concentration of the contaminants will decrease as the aquifer is cleaned up. The original design for the treatment system should include this change of concentration. We do not want to have to come back halfway through the project and put in new equipment.

FOULING

I will discuss one other area that often shows up with field installations. Do not forget fouling when designing an air stripper or a carbon adsorption treatment system.

Air Strippers

The main fouling problem occurring with air strippers is with iron. At neutral pH and in the presence of oxygen, soluble ferrous iron (Fe^{+2}) rapidly oxidizes to ferric iron (Fe^{+3}), which readily hydrolyzes to form the insoluble precipitate, ferric hydroxide, $Fe(OH)_3$. This precipitate attaches to the media of the air stripper. In addition, a bacterial slime usually forms in conjunction with the iron. We end up with lower efficiency with the air stripper, and high maintenance costs for cleaning the media on a periodic basis.

On small-scale systems I have used an aeration tank and filter before the air stripper to prevent this problem.

Fouling can also occur from hardness in the water and bacteria growing on organics in the water. It is very rare to have hardness be the cause of fouling in an air stripper. Iron can cause problems as low as 5 mg/L and below. Hardness must be about 500 to 1000 mg/L for a problem to occur. There is usually not enough evaporation in an air stripper to cause precipitation from hardness.

Air strippers can also be fouled by bacteria growing on the organics in the ground water. This will only happen with degradable organics. But, when stripping ring compounds (BTX, etc.) or ketones (MEK, MIBK, acetone, etc.), bacteria can readily grow on the media and reduce stripper efficiency. Any time that the degradable organics are above 10 mg/L this can be a problem. One of the best ways to solve this problem is to place a small biological treatment unit before the air stripper. Once again, suspended solids should be removed before the air stripper.

CARBON ADSORPTION

Fouling can also occur in carbon adsorption systems. Fouling in carbon causes back pressure, and the carbon has to be backwashed or replaced. Backwashing will ruin the adsorption zone in the carbon, and carbon is too expensive to replace due to solids buildup. Iron is the main culprit in carbon fouling, but bacteria can also be the cause. Pretreatment with aeration or biological treatment are the best solutions.

The important point to remember is that there is material other than the contaminant in the ground water that can affect the treatment system that you install. It is important to understand the treatment process and its limitations. Just because you know that you have 50 ppb of TCE, and that you want to create drinking water, does not mean that you are ready to design a treatment system.

REFERENCES
1. *Groundwater Monitoring and Sample Bias*, API publication no. 4367 (June 1983).
2. *Manual of Groundwater Sampling Procedures*, NWWA-EPA Series (1981).
3. *Practical Guide for Groundwater Sampling*, IL State Water Survey (Nov. 1985).

CHAPTER 14

The Effect of Time on Treatment Economics

Evan K. Nyer

This chapter will examine a single factor in comparing groundwater treatment systems — time. As we all know, we cannot simply take the initial concentration and flow rate, design two or three treatment systems, and compare them by capital cost. The capital cost is only one factor in the cost of a treatment system. All treatment systems have operational costs. These costs consist of power, personnel, supplies, sludge disposal, packing maintenance, carbon disposal, and replacement. To get the true cost of a treatment system, we add the capital and operating costs. (I realize that I am leaving out such costs as engineering, installation, demobilization, etc., but I want to keep this analysis simple.)

The one question that remains is, how long will the project last? Therefore, the point that I wish to discuss is the effect time has on the selection of a treatment system.

We will use three cases to study the effect of time. First, we will compare

ISBN 0-87371-731-7

small quantities of carbon to large quantities of carbon. Second, we will look at carbon adsorption vs air stripping, and third we will compare carbon adsorption to air stripping with vapor phase carbon. Finally, we will look at all of these comparisons using life-cycle design considerations.

CARBON VERSUS CARBON

Our design basis for comparing large quantities of carbon to small quantities of carbon will be:

- Flow — 40 gpm
- Benzene — 10 mg/L

There are no other contaminants present, and there is nothing that can interfere with the treatment process, i.e., silt or iron. These same assumptions will be maintained in all of the examples.

In Chapter 10 I discussed using small-size carbon units to treat low concentrations of BTXE. The carbon units used were 800-lb units, and had a capital cost of $4200. The operating costs mainly came from replacement carbon, and were listed at $1.25 per pound for new carbon, $0.75/lb for disposal, and $0.75/lb for transportation. Since that time I have worked on several projects; I now think that more accurate numbers for disposal and transportation would be $1.75 and $1.50, respectively. As always, the reader should check local conditions and develop his own specific pricing. For this analysis I will use a total of $4.50/lb for small quantities of carbon. A single carbon unit will be used, and I will assume a carbon usage rate of 8 lb of carbon per pound of benzene. For our design basis we would have a capital cost of $4200 and a daily operating cost of $173.

In Chapter 12 I discussed using a relatively large carbon adsorption system. That system consisted of two 2000-lb carbon units. The capital cost for that system was $45,000, but the carbon replacement costs were only $1.25/lb (new carbon plus regeneration of used carbon). I will assume a small change in the transportation cost and list a total cost for carbon of $1.75/lb. Since this system uses two-stage carbon, I will assume a carbon utilization of 6 lb carbon per pound benzene. The daily operating cost would be $50.40.

To portray a comparison of the costs of these two systems, Figure 1 depicts the cumulative costs for both systems. As can be seen, the break-even time is approximately 0.9 years. Therefore, the small carbon system should be used for projects less than 0.9 years and the large carbon system should be used for longer projects.

This is a pure operating vs capital cost comparison. By using the same technology I have kept the entire discussion to high capital costs with low

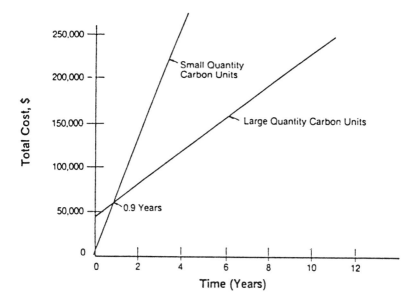

FIGURE 1. Cumulative costs — small vs large quantities of carbon.

operation costs vs low capital cost with high operations costs. Time is the deciding factor between the two situations. To get a full picture of this evaluation, we should redo the calculations at 5 mg/L benzene concentrations. You will have to trust me, the break-even point increases as you decrease the benzene concentration.

CARBON VERSUS AIR STRIPPING

Of course, the same type of comparison can be useful when comparing two different technologies. The design criteria for this example will be:

- Flow — 40 gpm
- Benzene — 1 mg/L

The rest of the assumptions will remain the same.

The small quantity carbon system will be used in this section due to the lower benzene concentrations. The capital cost will be $4200 and the operating cost will be $17.30 per day. In Chapter 10 I discussed using a small air stripper to treat BTXE. For this analysis we will assume that a 14-in diameter stripper will be sufficient for our design. The capital cost is $11,000 and the operating cost is $0.60 per day.

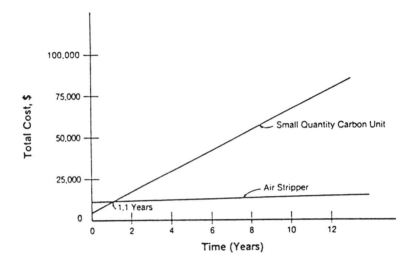

FIGURE 2. Cumulative costs — carbon adsorption vs air stripping.

Figure 2 compares the time effect on the costs of the two systems. The break-even point is 1.1 years. Once again, short-term projects should use carbon and long-term projects should use air stripping.

OK, everyone knows this. Stick with me, I may still have some surprises up my sleeve.

CARBON ADSORPTION VERSUS AIR STRIPPING WITH VAPOR PHASE CARBON

We can take the last comparison one step further by requiring that the air emissions from the air stripper are treated. All of the capital and operating costs for the carbon will remain the same. The air stripper will now require a vapor phase carbon unit and a heater. The heater is required to maintain the air at below water saturation. This is necessary in order to obtain higher carbon efficiency in an air stream than is possible in a water stream. Accordingly, the carbon capacity will be 2 lb carbon per pound benzene for the vapor phase carbon. Vapor phase carbon costs a little more than liquid phase carbon. Therefore, the total carbon costs will be increased to $5/lb for carbon. Finally, the heater will require 1.9 kw to increase the air temperature to required levels. The daily costs for the air blower, heater, and carbon will be $0.60, $4.50, and $4.80, respectively. The total daily cost for operation will be $9.90.

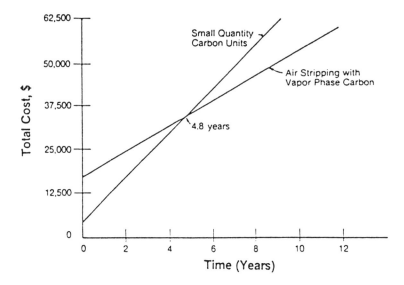

FIGURE 3. Cumulative costs — carbon adsorption vs air stripping with vapor phase carbon.

Figure 3 compares the cost of the two systems. As we see, it takes 4.8 years to break even. In other words, it takes 4.8 years for the savings from higher carbon capacity to overcome the initial capital expenditures. We will require a much longer project before we would switch to the higher capital solution.

LIFE-CYCLE DESIGN

I have made a point in all of my articles of doing designs by the "life-cycle" method. In the first part of this chapter I used constant concentration design methods. This would have been correct if we had been dealing with a leachate. Let us redo all of the preceding evaluations, but assume that the benzene concentration decreases by 50%/year. In other words, the initial concentration is 10 mg/L. At year one, the concentration is 5 mg/L, at year two the concentration is 2.5 mg/L, etc. To once again keep it simple, I have calculated the values at each year, then used curve fitting to project between the data points. Figure 4 depicts a life-cycle comparison of small quantities of carbon vs large quantities of carbon. There is a significant effect on the total costs for both technologies. However, the break-even point is still about the same time. Other than being very conservative in our budgets, the decision between technologies would remain the same.

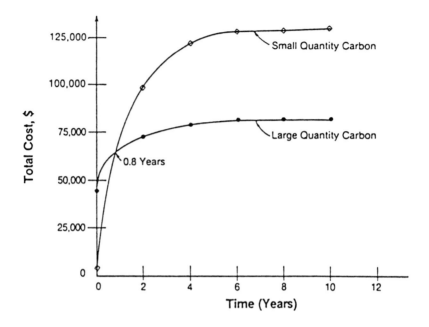

FIGURE 4. Life-cycle design — carbon vs carbon.

Figure 5 illustrates the life-cycle comparison for carbon vs air stripping. Life-cycle design does not affect the air stripping cost at all. Once again the total cost for the carbon is significantly lower. However, as we saw previously, the break-even point is about the same. Life-cycle design will not affect our selection of technology.

Figure 6 shows the life-cycle comparison for carbon vs air stripping with vapor phase carbon. Of course, this is my favorite figure. Life-cycle affects the total cost for both technologies. The surprise is that the curves never cross. We never recover our investment for the vapor phase carbon system. For this case, we not only will have more accurate values for our budgets, we will actually change our decision on the equipment used to treat the contaminated ground water.

As I have stated before, you should not use the exact numbers presented in this chapter. Develop costs that are associated with your specific project and reflect local conditions. The thing to remember is that time will have an effect on the technology that we use and that life-cycle design can have a surprising effect on the choice of technology.

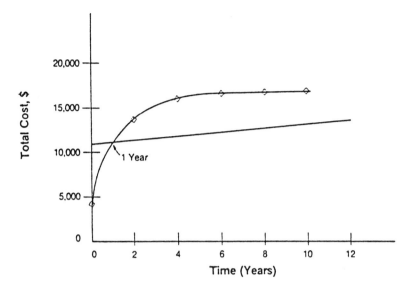

FIGURE 5. Life-cycle design — carbon vs air stripping.

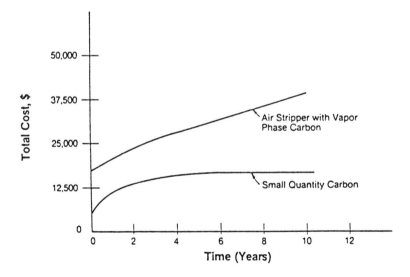

FIGURE 6. Life-cycle design — carbon vs air stripping with vapor phase carbon.

ACKNOWLEDGMENTS

I would like to thank Robert T. Ackart of Geraghty & Miller Engineers, Inc. for his assistance. He performed all of the calculations and developed the graphs.

CHAPTER 15

One Small Voice for Pump-and-Treat

Evan Nyer

Something happened during December 1989. All of a sudden, the work that I (and a lot of other people) had done during the last 10 years was wrong. Not just out of favor, but technically worthless. The EPA and the office of Technology Assessment (OTA) decided that pump-and-treat is no longer a viable technology. Both of these organizations decided that old is out and that a new technology will be found that will save us all from the "pollution demon". EPA does not know what that technology is (it may be biotechnology), but it is sure that it is out there. OTA basically wants to "nuke" any contamination. Some of the best quotes on this topic include:

- "EPA's Superfund office does not convey the generally negative view about pump-and-treat consistently found in the technical community". — OTA[1]
- "The large commitment of money to pump-and-treat groundwater clean-ups may be largely misdirected". — OTA[1]

ISBN 0-87371-731-7
© 1992 by Lewis Publishers

- Groundwater extraction may not "fulfill the primary goal of returning ground water to its beneficial uses". — EPA[2]
- "We desperately need technology breakthroughs". — Reilly[3]
- "Engineers want to stay with the tried and true. Well, the tried and true is costing too damn much". — Reilly[3]
- "If we have a magical torch, it's biotechnology research". — Reilly[3]
- "Grassroots activists that fight mobile incineration remedies are working against their own interest". — Hirshhorn, OTA[4]
- "The worst deal is bioremediation". — Hirshhorn, OTA[4]

To be perfectly fair, I have extracted the most sensational quotes here. The following quote is representative of the more responsible voices:

"Pump-and-treat groundwater remediation, while successful in containing contaminated groundwater plumes and reducing the concentration of groundwater contaminants, cannot be relied on to bring contaminant levels down to environmentally accepted standards". — Hanson, EPA[2]

The problem is that the news media will not pick up the responsible quotes and give them the same headlines that it will give to the sensational quotes. So, while the EPA has not exactly condemned pump-and-treat, you could easily conclude that from the headlines. I decided to go one step beyond the headlines, and I acquired the original reports that were the basis of the quotations.

Most of these quotes were the result of three reports. One is from the EPA and is titled, "Performance Evaluations of Pump-and-Treat Remediations".[5] The second is also from EPA and is titled, "Evaluation of Ground-Water Extraction Remedies".[6] The final report is from OTA and is titled, "Cleanups and Cleanup Technology".[7]

Joseph Keely wrote the first EPA report. It is an excellent review of pump-and-treat. The report concentrates on hydrogeology. It also describes how the anomalies of a particular site can greatly affect the efficacy of a pumping system. Basically, the report shows that aquifers are not sand boxes below ground. I would suggest that everyone get a copy of this report from the EPA and read it carefully.

The second report from EPA is more typical of its products. It is a good summary of the data that is available, but does not do a good job of interpreting that data. EPA's conclusions are based on a pure extrapolation of the data, and no insight into the real meaning is provided. I would recommend this report only if the reader wanted to review the information that formed the basis of some of the quotations that have surfaced lately.

The final report is from OTA. The most polite description of this report is that it is a diatribe. These people need to get out of Washington and see what the real world is like. A little technical education (I should be nice and say a "broadening" of their technical education) would probably not be a bad idea either. Unless you feel like getting mad, I do not suggest that you read this report. Unfortunately, most controversial quotations are taken from the OTA report. One thing stands out clearly, OTA believes the only "permanent" solution to contaminated sites is incineration.

After reading all of the news headlines and then reading the reports on which the quotations were based, I have some suggestions.

SHUT DOWN THE "AL HAIG" SCHOOL OF PUBLIC RELATIONS

Who is in charge here? The last time that I looked, we had one administrative branch of government. All of the people who are quoted report to that branch of the government. The only thing that they have collectively accomplished is to confuse everyone.

They have confused the technical community — Where are the hydrogeologists and engineers who are currently faced with designing cleanups to find information on the proper methods of design? What technical methods should be used? All of the methods that have been described in the technical literature have just been declared inappropriate. How are these technical people supposed to present a solution to a client that the government will accept?

They have confused the government employees — All of these reports are basically negative. They say that pump-and-treat systems are not viable, but do not provide an alternative. How is someone in the EPA supposed to review a proposal for a remediation? Should they not approve pump-and-treat proposals? How are the EPA contractors supposed to design their systems?

Most importantly, they have confused the public — What is the public to make of these headlines? It has basically been told that not only is Superfund slow, but that even when treatment systems have been installed they do not work. These reports, especially coming out as sensational headlines, take all public confidence away from Superfunds. I would like to remind the EPA and the OTA that it is the public who ultimately decide how much money will be spent on these cleanups. We need an understanding public in order to continue this vital work. The type of quotes presented here can only undermine the public's confidence, and remove its support.

My first suggestion is to educate, not confuse. The federal government needs to review the various technical opinions and formulate a single recommendation. It cannot allow groups within itself to have a public debate on these subjects. The federal government then needs to disseminate this information. The technical community needs to be kept informed on the latest technology and field applications of existing technology. The government employees (especially EPA and its contractors) need to be continually educated on the subjects that they are reviewing. The public needs to be educated on the real time and technology needed to complete cleanups.

WHO SAID THAT PUMP-AND-TREAT CAN COMPLETE A CLEANUP ON ITS OWN?

There are two basic problems here. First, who ever said pump-and-treat was a sole method of remediation? Second, cleanup as an "immediate" concept forces restrictive designs. Both of these areas must be addressed if we are to have effective remediation designs.

I have never met anyone who felt that a pump-and-treat system was the only process that was required to clean up a contaminated site. My articles and book have always stated that the concentration of the contaminant will reach an asymptotic level as the cleanup continues. People such as Richard Raymond have stated for 20 years that pump-and-treat systems by themselves cannot treat adsorbed organics in a reasonable time. How did the designers of EPA Superfund projects determine that purely pump-and-treat systems could clean a site?

There is a second basic problem with stating that pump-and-treat cannot clean up a site in a reasonable amount of time. The government has mandated that a Superfund project cannot be taken off the National Priorities List (NPL) until it is completely remediated. The public perceives that nothing has occurred until a Superfund site is taken off the NPL. This "Catch-22" process leads to the belief that pump-and-treat has failed because the site has not been taken off of the NPL. The government must understand and the public must be educated to the fact that complete cleanup, based upon the best technical and environmental criteria, may take extended periods of time. Another criteria must be found to measure the progress of Superfund cleanups. We must then let the public know that we are making progress and use a new method of review to show that progress rather than EPA's archaic "bean counting" system.

To make sure that I do not just criticize without making suggestions, I will quickly propose an alternative method to measure the progress of Superfund cleanups. I suggest that we use risk assessment as our basis for measuring the progress of site remediation. We could note that major steps (e.g., drum removal) would show a major reduction of the risk that site posed to the public. We would also be able to demonstrate when treatment of the remaining contaminants show a gradual lessening of the risk of the site. The risk would be measured in specific numbers so that we could calculate percent of cleanup, but we could also compare these numbers to risks that the public face every day. The public would then be able to understand the situation without a technical education. For example, we could compare the risk at a particular site to smoking a pack of cigarettes per day. We could then show that our remediation plan had lowered the risk to that of sun bathing. The public would be able to intelligently decide if spending an extra billion dollars was worth the reduced risk, and Congress could be told that it got a 98% reduction in risk for its $9 billion Superfund expenditures.

INCINERATION IS NOT THE ANSWER TO ALL REMEDIATIONS

Normally, I would limit my comments to ground water or to systems that have a direct effect on the cleanup of ground water. However, the OTA report was so negative in its attack on any system that was not incineration that I feel that a few comments are appropriate.

I disagree that incineration is a complete remediation process. There are residuals with any incineration process; the most obvious of which is ash. If you take soil and increase its temperature by a couple of thousand degrees, you basically destroy most of it. All of the organic material is destroyed, but the inorganic material remains. That inorganic portion contains all of the heavy metals that were previously present. Even if no specific heavy metals are present, the remaining material is worthless. Incineration produces ash, and you have to find a "permanent" place for that ash. My personal opinion is that I would rather have 5 ppb of toluene in the soil in my back yard than a pile of dead, black ash. The 5 ppb will continue to decrease with time. The ash will sit there forever.

Second, incineration produces air emissions. Now, we can argue the relative risk of soil contamination vs air pollution until the final Superfund site is cleaned up (i.e., forever). But, the basic question comes down to what does the public want? Try siting an incineration in your area and you will discover that the public does not like incineration. The OTA needs to take the pulse of America and stop trying to force incineration down the country's throat.

WHERE DOES PUMP-AND-TREAT FIT INTO REMEDIATION?

The only way to end this chapter is to note when and where pump-and-treat should be used. I firmly believe that we need to use pump-and-treat as we have always used it — to control plume movement. If you have a contaminated plume moving off-site, the proper way to control that plume is to pump in order to change the flow patterns in the aquifer, and to recover the water for treatment at the surface. Even when we want to use in situ technology, we pump the water in an aquifer to provide the required flow patterns across the contaminated zone. We treat the excess water that is not returned to the aquifer as part of the in situ program.

When we use pump-and-treat to remediate a site, a simple calculation can show the limitations of such a plan. Assume that an organic compound that we want to remove from the aquifer has a retardation factor of 20. Your homework for tonight is to calculate how many pore volumes it will take to move that compound to our well that is 50 ft away. Assuming a permeability of 10^{-5} cm/sec, how many years will it take to remove that compound from the aquifer? When you do the calculations, you have assumed that you have a perfect sand box below ground. Read Keely's report to find all of the things that will make the process even slower. You quickly can see that the remediation will take a long period of time.

The real question then is not if pump-and-treat works. The real question is do you want to take 20 years and spend $1 million to remediate a site or do you want to take one year and spend $20 million to remediate a site? With proper pump-and-treat, we can control the site and produce low risks during the entire process. This is both environmentally sound and economically reasonable.

How about all of these questions that the EPA reports brought up? Keely's report is the best, and it has raised several good points. However, the main point seems to be that we need to use experienced hydrogeologists when designing groundwater systems. We cannot simply poke holes in the ground and pump. Aquifers are complicated, and its takes years of experience to understand these systems. To simply state that these systems are complicated and, therefore, we are not capable of designing proper pumping programs, is a slap in the face to every hydrogeologist in the world.

REFERENCES

1. "EPA Recommends New Litigations on Pump-and-Treat Cleanup Techniques," *Groundwater Pollution News.* 2: (10) (Dec. 1989).
2. "Pump and Treat not Effective Enough: EPA." Barron, T. WTN Special Superfund coverage paper (1989).
3. Superfund: the Stuff Myths are Made of. *Civ. Eng. News Mag.* (Jan. 1990).
4. *"Activists Urged to Push Incineration."* from Regulatory Briefs — Superfund paper, Pasha Publications (Dec. 4, 1989).
5. "Performance Evaluations of Pump-and-Treat Remediations." Keely, J. F. EPA Ground Water Issue, EPA/540/4-89/005 (Oct. 1989).
6. "Evaluation of Ground-Water Extraction Remedies," Vol 1, Summary Report. EPA Superfund paper EPA/540/2-89/054 (Sept. 1989).
7. "Cleanups and Cleanup Technology" — OTA'S Coming Clean paper (Oct. 1989).

CHAPTER 16

The Five Worst Design Mistakes I Have Seen on Superfund Projects

As discussed in Chapter 15, Superfund projects have become an easy target for anyone to take shots. The Office of Technology Assessment (OTA) has been having a field day with the data that is being produced on the current treatment systems. The Environmental Protection Agency (EPA) has also been reviewing the Superfund program. Every congressman who wants to show that he is an environmentalist, finds Superfund an easy target. Newspapers and magazines have also found Superfund to be defenseless.

Well, now I guess it is my turn. However, instead of attacking Superfund by the normal means, I plan to use the technical approach. Everyone complains that Superfund has not been able to meet its goals. Too few sites have been cleaned up. Too much money has been spent. I disagree with most of these complaints. I personally feel that the main problem with the Superfund

ISBN 0-87371-731-7
© 1992 by Lewis Publishers

program is that a bad job has been done of educating the public as to what can be accomplished and in what time frame. In this situation, it has been very easy for any group to take pot shots at the program.

However, the purpose of this book is to provide information on treatment of ground water. So, instead of reviewing all the problems with Superfund, I would like to concentrate my review on the technical problems that I have seen with the Superfund program. I have had many opportunities to review Superfund designs that were performed by EPA contractors. In general, I have been disappointed with the quality of these designs. In all the following examples, I have changed some of the design numbers in order to ensure that no one could directly relate the example to a specific design in the field. The pertinent data is correct, but the specific numbers have been changed (to protect the innocent).

THE WORST WELL DATA

Table 1 lists the data taken from five monitoring wells. I realize that data from a Superfund site is never this simple. There are multiple samples from usually more than five wells. We have to deal with a range of concentrations from each well and a possible time effect on the values. I have simplified the data because of limited space.

The designer took the data presented in Table 1, developed design criteria, and performed a feasibility study. The results of that feasibility study were the selection of a treatment system to remove the compounds found in the ground water. Table 2 is a summary of the design criteria. As noted, both the average and maximum concentrations of each component were listed in the design criteria. All of these averages assume equal weight to the data points. This may not be an accurate assumption. If the area represented by one well contributes most of the flow, then the data from that well should be weighted higher in the averages. Once again I have had to keep it simple here.

The organics listed in the table can be treated several different ways. Air stripping was selected as the technology. I am going to concentrate on the pretreatment requirements of this ground water. Pretreatment of the suspended solids is required to keep the maintenance on the air stripper reasonable. Also, there are limitations on effluent suspended solids. The flaw in this design occurred with the pretreatment system.

From groundwater modelling, the flow rate was determined to be 200 gpm. This is the rate that was required to completely capture the plume. The designer decided to use the average concentration in the pretreatment system, which was then checked for performance under maximum concentrations.

The analysis of the data from the wells determined that the main criteria

Table 1. Monitoring Well Data

	MW 1	MW 2	MW 3	MW 4	MW 5
Volatile organics, ppb					
Trichloroethene	30	80	40	100	10
1,2 Dichloroethene	15	30	5	55	4
General water criteria					
pH	6.6	6.7	6.5	6.7	6.8
Temp, °F	52	53	52	53	52
Fe, ppm	0.5	0.7	1.0	0.4	0.8
Ca, ppm	40	44	48	44	42
Suspended solids, ppm	480	1000	20	25	1100

Table 2. Design Criteria

	Average Concentration	Maximum Concentration
Volatile organics, ppb		
Trichloroethene	52	100
1,2 Dichloroethene	22	55
General water criteria		
pH	6.5–6.8	
Temp, °F	52–53	
Fe, ppm	0.7	1.0
Ca, ppm	44	48
Suspended solids, ppm	525	1100

for the pretreatment design would be the suspended solids removal from the ground water before it reached the rest of the organic treatment system. The system was designed for iron removal by aeration and then polymer addition, flocculation, and clarification for the suspended solids removal. Figure 1 shows the design for this treatment system. You will note that due to the large amount of suspended solids, they also were required to install thickeners and filter presses. Based on design data, this plant should produce 1,300 lb of dry solids per day for disposal. Assuming a 35% cake can be produced by the filter press, about 21 yd^3 of solids will have to be disposed of on a daily basis. That means at least one truckload per day.

What is right with this design? The designers included aeration for the iron before the clarification. Also included was the correct clarification and provision for the possible need of polymers. They included the disposal of the solids from the system, not just the removal. What did they miss?

The answer is quite simple. Whoever heard of a well producing 500 mg/L of suspended solids. It would be obvious to an experienced hydrogeologist that the monitoring wells used for the production of this data were not properly developed. This is a case in point in which there is adequate design performed, however, the lack of experienced personnel in reviewing the data forced a relatively simple treatment system to be designed as a very complicated, and expensive treatment system. Experience would have shown that the data was erroneous and that no well could produce 500 mg/L suspended solids if it was developed properly. Experience would have been able to save the Superfund project a significant amount of money. A better technology would not have helped in this case.

THE WORST LABORATORY DATA

The next example was taken from a design for iron removal. Representative samples were taken from the ground water and delivered to a laboratory for the study. The purpose of the study was to test the various iron removal techniques to determine which would be best for the full-scale plant. One of the nice things about inorganic removal is that laboratory data is usually accurate enough to design full-scale systems from it. Normally, there is no need to perform pilot plants to confirm the data. When designing an inorganic treatment system there are several basic questions that must be answered. They usually include:

- Dosage of chemical for optimum precipitation
- Optimum pH for precipitation
- Concentration of suspended solids produced from precipitation
- Requirements for polymer or other coagulant aid
- Other data, depending on the specific inorganic to be removed by the system.

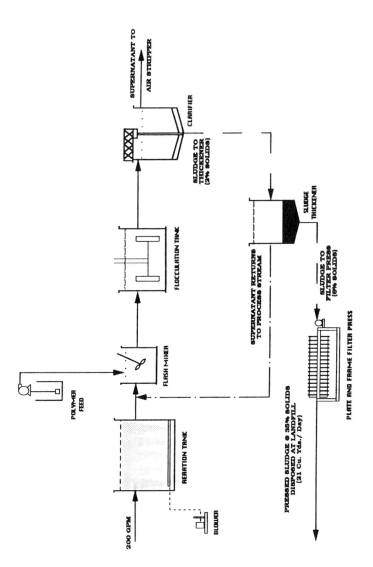

FIGURE 1. Pretreatment system.

This data, combined with experience with flocculation and clarification, can lead to a direct full-scale design.

The problem is that we must rely on laboratory data to produce the design data that we need. Most of the time the designer receives data summarized in tabular form to use on his design. We cannot simply treat this data as absolute. We must understand how the data was generated and its limitations. Some of the worst designs can be based on false data. Usually we can get the entire Feasibility Study (FS) before a design is done in detail. When laboratory work has been included as part of the FS, we need to read the methods used in the laboratory, not just the data summarized from the treatability study. In our present example, the laboratory treatability study was performed to determine the optimum method of removing iron from the water. Five methods were tested: alum, ferric chloride, sodium hydroxide, lime, and aeration. All these chemicals (oxygen from the aeration) are capable of precipitating iron from water. The idea of the laboratory study was to find the optimum method for removal.

The laboratory work had immediate problems in that they waited for the samples to reach the lab to measure temperature, pH, and oxygen levels in the water. It is absolutely impossible to have accurate measurements of these three parameters in samples that have been brought back to the laboratory. These parameters must be measured in the field and, preferably, in the well itself.

The next mistake of the laboratory treatability study was not to optimize each chemical addition. The proper method of studying chemical precipitation is to optimize each chemical addition and to optimize the pH at which it performs best. The problem is that a tremendous number of tests are required to come up with the optimum of each chemical addition and then develop an optimum between the different chemical methods. The treatability study took a short cut and simply used the literature to determine chemical dosages. The problem with this thinking is that it does not produce an accurate comparison between chemicals and does not develop accurate data on the amount of chemicals required or the suspended solids produced. These two data points are critical to the design of the full-scale system. In fact, solids disposal may be the most costly item on a long-term basis.

The basis of the study was to add the chemical, do a visual inspection of the floc that was formed from the chemical addition, measure the resulting suspended solids created, and measure the iron left in the supernatant. The real flaw in the study came at this point. All the other problems I mentioned before would have made the plant slightly less cost effective, but would not have made this study totally invalid. The visual inspection showed that the alum, ferric chloride, sodium hydroxide, and lime all produce a large amount of floc that settled readily. The aeration system, however, only produced a very small amount of floc that did not settle readily. The researchers took this visual data to mean that the aeration system failed; they did not continue the aeration test at all. They proceeded to measure the suspended solids produced and the result in iron concentration in the supernatant from the alum, ferric chloride, sodium hydroxide, and lime only. The aeration study was stopped at this point in time. No measurements were taken, and the iron sample was, literally, thrown out.

What these researchers failed to realize was that the lack of large floc generation should have been considered a success, not a failure. Large floc production directly relates to large amounts of suspended solids. The increased floc was due to the chemicals that were added. Aeration added no extra chemicals. Only the iron was in the floc that was formed.

As stated before, the most expensive part of an inorganic treatment system is often the removal and the disposal of suspended solids. Any precipitation step should be optimized for low production of suspended solids. Small and slow settling floc can be solved by other methods. Polymer addition to produce larger floc and a lamella clarifier are two possible methods to overcome small floc conditions. The researchers would have known this if they had simply continued on their original experimental design. The problem was that upon visual inspection of the iron floc, they aborted that test. If the test would have continued they would have shown that the main goal, iron removal, was accomplished by the simple aeration.

All these procedures and methods (mistakes and all) were clearly stated in the feasibility study report. I do not understand where the reviewers from the contractors and the EPA were when they read this report. They too had to accept this data as a basis for a full-scale design. The designers went on to base their treatment system design on the high concentration of suspended solids generated from one of the chemical addition methods. Once again, the system became controlled by suspended solids removal, thickening, and de-watering.

This is another case-in-point where new technology would not have helped reduce the cost for subsequent generation of large volumes of hazardous solids. Once again, experience, knowledge of the significance of the data, and the significance of what that data produced in the design were what was needed. The reviewers needed the experience to understand what a full-scale system would entail and the experience to know that the laboratory procedures were flawed. Experience, not new technology, would have shown cost savings on this project.

THE WORST FLOW SELECTION

Very mundane parts of the design can sometimes rise up to create an inefficient design. One of the strangest design bases that I have come across is in flow selection. Flow selection is normally tough to determine for a full-scale treatment system. Problems with inaccurate data on wells, aquifer modelling, draw down effects on multiple wells, degeneration of wells because of iron or suspended solids or bacteria can all combine to create a situation in which original estimates for flow are not what the final treatment system receives. All of these must be taken into account and the best estimates must be produced for the final treatment system. You cannot assume that simply because a system is designed for higher flow rates that it will be a conservative design and that any smaller flow rate will simply be treated more efficiently in the larger system. This does not make economic sense nor does it make technical sense with several technologies. There are several units that are designed based on residence time; significantly going over those residence times can ruin the treatment system.

The worst example that I have seen for flow selection is a case in which the data from the monitoring wells and modelling showed that 11 gpm would be required to completely capture the plume. The designers took this data and through various manipulations finished with a design basis of 80 gpm for the treatment system.

Several strange assumptions were required to get from the original 11 gpm to the 80 gpm design. First, the designers decided that the most expensive part of the operation would be operator attention. Normally, a good assumption. They assumed that any treatment system would require around-the-clock operator attention and that this would be too expensive for a small treatment system. Their first manipulation, therefore, was to determine that they only wanted to run on an 8-h shift for the treatment system. Then they decided that by the time the workers got to work and by the time they cleaned up at night the system would only run 6.6 h/d. Based upon this logic, they ended up with a 40 gpm treatment system with a storage tank to collect the overnight flow. One final manipulation jumped this from 40 gpm to 80 gpm. The designers decided that they would have to be conservative and design the system for potential doubling of the flow. It was never stated whether this assumption was based upon the wells producing twice the flow they originally calculated or the workers working half the time.

Basically, this is ridiculous. Not all treatment systems require around-the-clock operator attention. Any good designer can limit the operator attention required on a treatment system. This is especially true for small, 11 gpm treatment requirements. I have personally designed biological treatment systems, which are considered very operator intensive, for abandoned sites. Even biological systems can be designed to only require attention once a week or once every other week.

Taking the design from 11 gpm to 80 gpm also severely restricts the technologies and the specific designs that can be used. For example, air stripping technologies at 11 gpm have a full range of choices between diffused aeration and standard packed towers. An 11 gpm system is small enough so that diffused aeration systems do not become too costly. At 80 gpm, on the other hand, packed towers are favored. In addition, the simple cost increases from increasing the size of the treatment system eightfold are evident. There will be increases in capital and operating cost for this type of size increase.

Once again it is not new technology that will save money in this case. It was a lack of understanding the technologies and the design that made the system cost too much. Experience, not new technology, was needed here.

THE WORST IRON REMOVAL METHOD

Any idiot can design a $20 million treatment system. Anyone can research in books and articles all the possible methods that can be used to treat a specific compound. Then all those methods can be put in series, thus assuring that the compound will be removed from the water. When money is no object, this method will assure success. If our only objective is successfully removing the compound, then any idiot can design a treatment system.

However, when money becomes a part of the formula for a successful design, we then require something more sophisticated than simply picking out a technology from the literature. It is amazing that we are allowing inexperienced designers to simply use literature as their base of knowledge. Without experience, one cannot fully absorb or understand what is written in publications on the specific treatment systems.

The next example goes back to iron removal. During the FS the designers decided to use green sand filters as the method for removing iron from the ground water. Now this becomes a tricky point. Are green sand filters a bad technology? No. Green sand filters are an excellent technology that have been widely used for iron and manganese removal. Then, how can I say that I consider the selection of green sand filters for groundwater treatment system to be one of the ''worst'' designs that I have seen?

For most groundwater situations, green sand filters would not be the best application. The problem is, how does a designer pick between the different technologies that are available to him for the particular removal method? The answer is, once again, experience. With experience you realize that green sand filters are difficult to operate. An oxidant must be used to regenerate the green sand properties and the filters must be backwashed to remove suspended solids. Significant amounts of sludge are produced for final disposal. Green sand filters have a high capital cost. A small groundwater system would require high automation which would translate into high cost for controls and electronics.

The trouble with green sand filters vs other iron removal technologies is that the best technology does not stand up and shout. All of the iron removal technologies have their place. Different water requirements can make any of the technologies the preferred technology for that particular water. It takes years of operational experience to understand the nuances between the different use criteria. And, again, we have all the technologies that we need for iron removal. What we do need is more experienced designers who apply these technologies to ground water.

THE WORST DESIGN METHOD

The final example is not a specific design. However, I have included it here because "The Five Worst Designs..." sounds better than "The Four Worst Designs..." in the title of this chapter. Superfund is currently using a system in which the low bid contractor is handed design responsibility. The best way to get a low bid is to use less expensive labor on the contract. The best way to get less expensive personnel is to use people that have just gotten out of college. It is not that these people are not intelligent or do not care about the designs that they are not performing. The problem is that these young professionals have not had sufficient time to gain the experience necessary to do adequate designs. In and of itself, this is not a huge problem. A lot of the work on an FS can be performed by well-trained, young professionals. The system has fallen apart because we have failed to use experienced designers when it comes to final process design and review of projects. In the Superfund designs that I have reviewed for our private clients I have seen a severe need for better review of the projects. I have listed several of the worst examples here. Others, of a more minor nature, occur all the time in these reports. I realize the entire system cannot be changed. However, I would suggest that the EPA change their internal review methods. I would suggest that reviews be done by a second contractor on Superfund designs. I would also suggest that somehow these second contractors not be hired by low bid but by a method that favors design experience.

I have used simple examples. As we get into more difficult treatment situations, new technology will be needed to keep the cost of treatment reasonable. But, even with new technology, experience will be required. I am currently working on a project where the EPA contractor picked a specific method for organic removal. One of the PRPs requested that biological treatment be considered instead. They provided data from a groundwater operation at one of their facilities. The contractor rejected the idea and never evaluated the option. The contractor's position was that biological treatment could not be applied to low concentrations of organic. They lacked the experience in this new technology and no textbooks were available on this new method.

This will end my two chapters on Superfund bashing. I realize that these are all very difficult problems and that the people in charge of the EPA are trying to the best of their abilities. However, I feel that they have concentrated too heavily on new technology as their savior. They need to realize that good solid engineering practices could go a long way in reducing the cost of treatment and improving the quality of treatment designs. I hope that the specific examples provided here will add to the experience of the readers.

SECTION VIII

In Situ and Natural Biochemical Remediations

The most exciting technology currently being applied to remediation of ground water and soil is biological treatment. The main problem with the application of this technology is the same as that being experienced with all treatment technologies for groundwater and soil remediations — lack of experience. While there is a wealth of information on the specific reactions that bacteria and other microorganisms can have with organic contaminants, there are very few installations that have been running for several years.

A second problem is that the experts in this field come at the problem from completely different directions. It is difficult to know which approach is correct. Microbiologists tend to concentrate on specific microorganisms. The trouble with this direction is that it can cost $70,000 to name a bacteria; I have never seen any evidence that knowing the name of a bacteria will help in the field when you are trying to degrade a hazardous compound. Their work is focused on the bacteria with very little attention to the environment in which the bacteria live.

University professors tend to think that the laboratory can simulate any activity that will occur in the field. As we discussed in Section III, the laboratory can be a powerful tool, but full-scale systems cannot always be scaled down to laboratory-size systems. They are focused on laboratory studies with very little field work.

Government employees tend to want to have absolute proof before a biological project is accepted. The problem with absolute proof is that it can cost an infinite amount of money. Their work seems to get bogged down with proof and simple results are not accepted without absolute proof that the bacteria were responsible for the removal of the organics.

Engineers, hydrogeologists, etc., all tend to relate the process to their own expertise or resources. Everyone wants to bring the process to the site, and everyone wants to use advanced technology. Genetically engineered bacteria are the most probable advanced technology in the near future.

This section expresses my feelings that biological treatment should be applied to remediations. Biological reactions are natural. We do not bring the process to the field; we simply enhance the natural reactions that are already ongoing. All of the expertise listed before, working in concert, are required for a successful biological treatment project.

We need to stop trying to do research that will make good headlines in a magazine or newspaper. We should be directing our efforts to understanding what is already going on naturally and how we can measure that activity through analytical tests. That knowledge would allow us to make the biological process more effective. Hopefully, the reader will use the chapters in this section to begin their understanding of the natural biological process.

CHAPTER 17

Priming the Pump for In Situ Treatment

Evan K. Nyer

In situ treatment is one of today's most exciting treatment concepts in the environmental field. For aquifers and soils contaminated with organic material, in situ treatment offers a natural destruction process at a relatively low cost. Industry, treatment experts, and governmental agencies are all embracing the concept as an important tool for aquifer restoration.

While everyone is excited about in situ treatment, and several cleanups are currently using this technology, there are two main roadblocks preventing widespread implementation. The first roadblock is one common with other treatment technologies. What is clean?

We need to establish organic concentration levels that are considered safe and acceptable to the public. The federal and state environmental agencies are already working on this problem. While we all appreciate the complexity of establishing specific concentrations, this work must proceed with diligence and haste. In addition, most of the work so far has concentrated on safe

ISBN 0-87371-731-7
© 1992 by Lewis Publishers

concentrations for water. Contaminated soils do not have to be as clean as water, but they are also a significant problem today. Safe and acceptable concentration levels must be developed for soils.

The second roadblock is unique to in situ treatment systems. When can we stop active management of an in situ treatment cleanup? We must not minimize the "what is clean?" question but, for in situ treatment, active management is a more important question at the present time.

Ground water cleanups are different from most cleanups that have been performed in the past. When a river or a lake has to be cleaned up, the main method is to remove the sources of contamination and the body of water cleans itself. One could argue that the river or lake uses in situ treatment since the contaminants are removed from the water without moving the water outside its normal boundaries.

Aquifers and subsurface soils do not clean themselves after removal of the source of contamination. While removal of the source of contaminant should always be the first step, the aquifer itself must then be cleaned.

What makes aquifers unique when we are trying to remove contaminants from water? First, almost all in situ cleanups are biological. Biological treatment systems require certain basic environmental conditions in order to work. They are:

- Bacteria that use the contaminant as a food source
- Oxygen (for aerobic degradation)
- Nutrients (ammonia, phosphorous, and micronutrients)
- Temperature
- PH

Surface bodies of water can supply all of these environmental conditions. Our rivers and lakes were polluted because we were putting in contaminants faster than the surface bodies of water could replace oxygen and nutrients. Once we removed the source of contaminant, the surface body of water could catch up.

Aquifers cannot always meet the environmental conditions required for biological treatment. The key to the design of an in situ treatment for an aquifer is to supply the bacteria, oxygen, nutrients, and other environmental conditions that are required for biological action. In situ treatment could be defined as "supplying the environmental components required for biological action on the contaminants present in the aquifer".

This brings us back to active management of an in situ program. Active management supplies the environmental requirements for the aquifer. This is done through the use of wells, pumps, chemicals, aboveground equipment, etc. All of the costs associated with in situ treatment come from the active management of the project. In order for in situ treatment to be a low cost method of cleanup, we must be able to define when we can stop the active management.

This is a different question than ''what is clean''. Aquifers have a limited ability to supply and replace their environmental requirements. Most groundwater contaminations exceed these abilities by orders of magnitude. However, once we have satisfied most of the requirements of in situ treatment techniques, will active management continue to have an effect on the removal of the contaminants from the aquifer? If the aquifer has sufficient supplies of bacteria, oxygen, nutrients, etc. required to satisfy the demand from the contaminants present, do we need to continue to have personnel and equipment on site? Or will the aquifer behave like a river or a lake and complete the cleanup by itself?

As this example shows, active management and ''what is clean'' are both important to aquifer restoration. Let us assume that 1 ppb of benzene is determined to be safe and acceptable in an aquifer. Let us also assume that at 10 ppb benzene, active management does not significantly increase the speed of the cleanup. Then active management could be stopped at the 10 ppb level, but the aquifer would not be available for use until it reached the 1 ppb level. (We did not go swimming in the lake the day after the local chemical plant installed a treatment system; rather, we waited until the lake cleaned itself of the contaminants present.)

These two roadblocks must be addressed before in situ treatment can be broadly applied. First, we must satisfy the public that the aquifers are safe. Second, we must be able to define when we can stop spending money on the cleanup.

As a final point, some readers may question why I have not discussed technology development as a possible roadblock. I do not consider technology development a roadblock to the application of in situ treatment. There are tremendous potential cost advantages with in situ treatment. The economics will provide the incentives for the private sector to develop the necessary technology in a timely manner. However, it will be up to the regulatory agencies to set the cleanup criteria for in situ techniques. These agencies must develop the final design criteria in a timely manner in order for in situ treatment to be broadly applied to aquifer restorations.

CHAPTER 18

Biochemical Effects on Contaminants' Fate and Transport

Evan Nyer, Victoria Kramer, and Nicholas Valkenburg

INTRODUCTION

One of the most important parts of a site investigation and remediation is the knowledge and understanding of the organic contaminants in the aquifer. We must understand the present location and the rate of movement of each organic. Most organic contaminants interact with the soil and do not move at the same rate as water does through the soil matrix. We refer to this phenomena as retardation; there are retardation factors for most organic compounds. Based on groundwater modelling and retardation factors, we can model the fate and transport of the organic compounds at a contamination site. We are then able to design a remediation system that encompasses all of the compounds at their present and future locations.

ISBN 0-87371-731-7
© 1992 by Lewis Publishers

While most fate and transport models are considered relatively sophisticated, recent investigations have shown these models to be limited. At the Seymour Superfund Site we have been able to actually follow the fate and transport of specific organic chemicals over the past five years. The original fate and transport models used were shown not to be adequate for a total description of the movement of all of the compounds. Basically, we found that biochemical reactions, in addition to retardation, have a major effect on the fate and transport of the compounds in an aquifer.

By reviewing the movement of five compounds, 1-4 dioxane, chloroethane, tetrahydrofuran (THF), benzene, and phenol, we can delineate the effect of the retardation factor and show the added effect of the biochemical reactions on the movement of these compounds. Studying the fate and transport of these compounds provides insight on biochemical effects and retardation factors as they range from nondegradable to easily degradable and very low to high retardation factors. This chapter should show that biochemical reactions will have to be considered in future work if a complete understanding of the movement of the organic chemicals is to be included in the design of the remediation program.

SOLUTE TRANSPORT MODELS

Most solute transport models in use today are a combination of groundwater flow models, which describe a velocity field, recharge and discharge, and changes in aquifer storage for site specific conditions, and transport models which may address dispersion, density-dependent flow, or some reversible and irreversible chemical phenomena. Most solute transport models do not address biochemical reactions. Reversible equilibrium and controlled sorption is typically simulated by use of a retardation factor or coefficient. Retardation is the dynamic process of adsorption to and desorption from aquifer materials and is compound specific. If a compound is strongly adsorbed it is highly retarded.

A retardation factor (R) must be calculated for each contaminant using the compound's organic carbon partition coefficient (K_{oc}), bulk density (B_d), and porosity (n) of the medium through which the contaminant is moving. Organic carbon partition coefficients can be calculated using empirical values of the octanol/water partition coefficients determined in the laboratory. The organic carbon fraction can also be determined in the laboratory. The organic carbon fraction was determined by laboratory analysis for the Seymour site. It could also have been estimated from the literature. Bulk densities are usually estimated from a knowledge of the grain-size distribution. Retardation factors for linear sorption are calculated using the following formulas:

$$R = 1 + (B_d K_d)n$$

where

$$K_d = K_{oc} \times \text{(organic carbon fraction)}$$

The retardation factor is a measure of how fast a compound moves relative to ground water. For example, a retardation factor of two (2) indicates that the compound is traveling at one-half the groundwater flow rate.

Dispersion, which is velocity dependent but not compound specific, requires values of hydraulic conductivity, hydraulic gradient, effective porosity, and dispersivity. The first two factors are usually measured in the field but the other two are generally estimated from the literature. Dispersion does not affect the migration rate of one contaminant compared to another, but does affect the concentration distribution in a contaminant plume.

DESCRIPTION OF THE SEYMOUR SITE

The Seymour site is located in Seymour, Indiana (Figure 1) and is 13 acres in area. The facility operated from 1971 to 1980 when the U.S. Environmental Protection Agency (USEPA) shut down the site because of releases to the air and surface water. After approximately 60,000, 55-gal drums and 100 tanks were removed by USEPA as part of its emergency response, the agency employed a subcontractor to conduct a Remedial Investigation and Feasibility Study (RI/FS) which focused on defining the extent of soil and groundwater contaminant concentration at the site. During the remedial design, additional soil and groundwater studies were conducted which resulted in a significant redefinition of the groundwater plume. Recent groundwater studies also indicate that the concentrations of contaminants have decreased with time and distance traveled from the site.

The results of the RI/FS and the remedial design work indicated that the soil is contaminated with volatile and semivolatile compounds and that a plume of contaminated ground water is migrating away from the site in the shallow aquifer in a northwesterly direction (Figure 2). The remedial program consists of:

- A RCRA cap to prevent further leaching of residual contaminants in the soil
- A vapor extraction system in the unsaturated zone to remove volatile organic compounds
- The addition of nutrients to the soil in order to stimulate the bioremediation of the semivolatile compounds
- A pumping and treating system to capture and remediate the contaminated ground water in the shallow aquifer

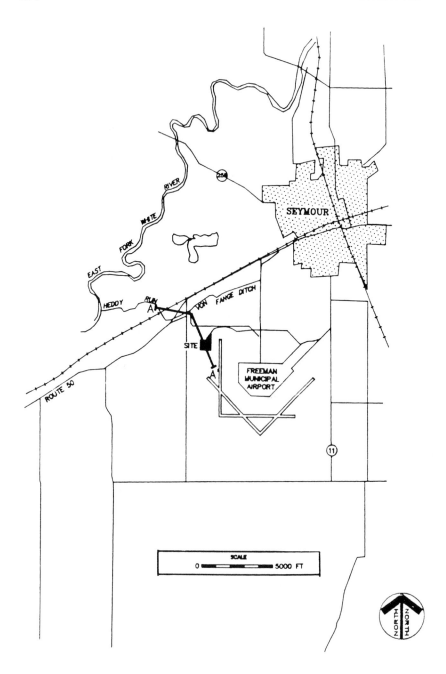

FIGURE 1. Site location.

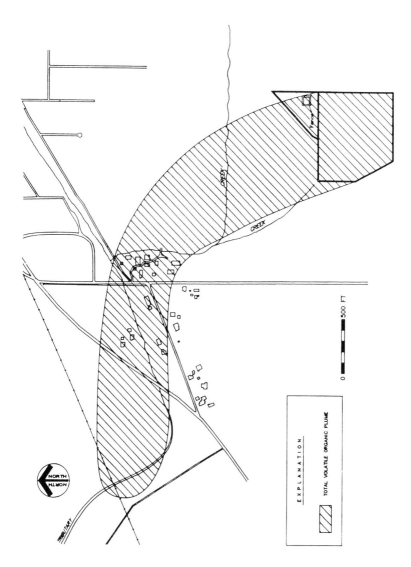

FIGURE 2. Extent of total volatile organic plume as defined in June 1990.

REMEDIAL DESIGN MODELLING

The main objectives of the design model for the Seymour site were:

1. To design and optimize the groundwater recovery system at the Seymour site
2. To assess the risk to the environment from operating the remedial systems
3. To estimate cleanup times and costs for operation and maintenance

To satisfy all of these requirements, a transport model was employed.[1] The flow model code utilized was the MODFLOW code.[2] The flow model was calibrated to observed groundwater conditions recorded at the Seymour site. Simulated flow conditions comprise the base of information which then became input to the transport model. The SWIFT code[3] was chosen as the solute transport model code. Only retardation and dispersion are accounted for in the model. Biochemical reactions are ignored in this transport model. The output from the model predicts the concentrations of compounds over time in the remedial pumping wells.

A recent evaluation of the actual concentrations found in ground water and expected transport based on retardation indicate that the model is likely to be overly conservative in its estimation of the transport of some compounds, particularly the compounds subject to biodegradation. Table 1 gives a list of some of the compounds in the groundwater plume and compares the distance each should theoretically have traveled, assuming entrance into the groundwater system in 1980, and that retardation is the only process that affects migration in the subsurface. The table compares the theoretical travel distances to those that have actually occurred, as determined by analytical data from monitoring wells at the site. Figures 3 through 7 are graphical representations of the data in Table 1.

Except for 1,4-dioxane, which has traveled almost as far as expected, the other compounds in Table 1 have migrated smaller distances than anticipated. Tetrahydrofuran and 1,4-dioxane, which have identical retardation factors and similar solubilities in water, exhibit greatly different migration rates. 1,4-dioxane has traveled about 2.5 times farther than tetrahydrofuran. This may be the result of the differing degradation rates of the two different compounds. Tetrahydrofuran is amenable to biodegradation (although at a slow rate), whereas 1,4-dioxane is not. In addition, benzene and phenol, which are also amenable to biodegradation, should have traveled 1,250 and 2,500 ft, respectively, beyond the site boundary. The analytical data indicate that these two compounds have migrated less than 100 ft from the site boundary. They are readily degraded.

The actual degradation rate of these five compounds depends on many factors.

● Are there bacteria present that recognize the compounds as a food and energy source?

Table 1. Comparison of Theoretical and Actual Contaminant Migration

	Retardation Factor	Theoretical Travel Distance[a] Beyond Site Boundary (ft)	Actual Travel Distance[b] Beyond Site Boundary (ft)
Chloroethane	2.1	2100	900
Tetrahydrofuran	1.0	4300	1450
1,4-Dioxane	1.0	4300	3900
Benzene	3.5	1250	100
Phenol	1.7	2500	—

[a] Assumes contaminants entered groundwater system in 1980 and retardation is the only process which attenuates the migration of each compound. Ground water in the shallow aquifer travels at 1.2 ft/d.

[b] From groundwater samples taken at the site.

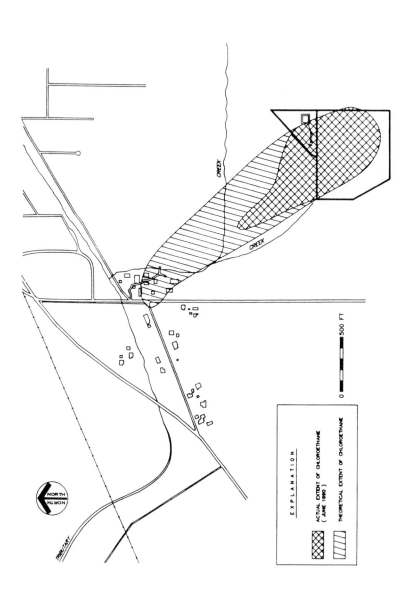

FIGURE 3. Theoretical and actual distributions of Chloroethane in the shallow aquifer, Seymour site, Seymour, Indiana.

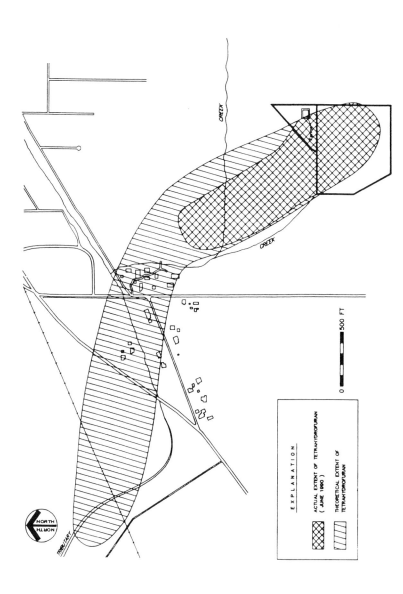

FIGURE 4. Theoretical and actual distributions of Tetrahydrofuran in the shallow aquifer, Seymour site, Seymour, Indiana.

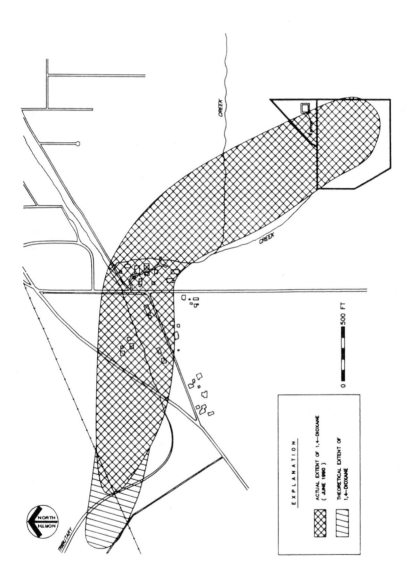

FIGURE 5. Theoretical and actual distributions of 1,4-Dioxane in the shallow aquifer, Seymour site, Seymour, Indiana.

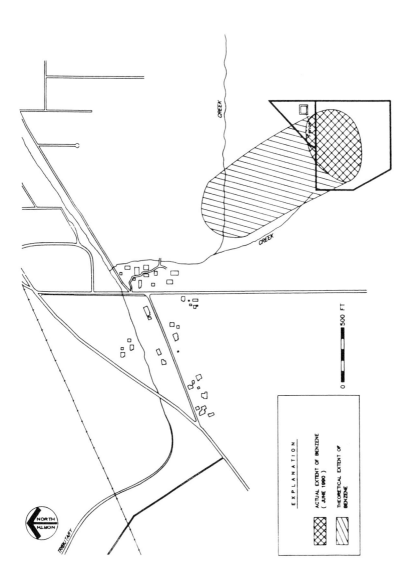

FIGURE 6. Theoretical and actual distributions of Benzene in the shallow aquifer, Seymour site, Seymour, Indiana.

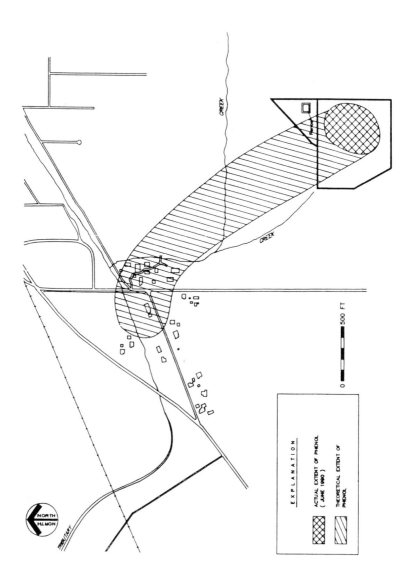

FIGURE 7. Theoretical and actual distributions of Phenol in the shallow aquifer, Seymour site, Seymour, Indiana.

- Are there co-metabolites present for the nondegradable compounds?
- Are the environmental conditions optimum for bacterial activity?
- Are there nutrients present?
- How much oxygen is naturally present for aerobic degradation?

We do not have sufficient data to determine if all of the factors are at an optimum for bacterial degradation. We also do not have direct evidence of bacterial degradation of the specific compounds at this site.

However, we find that the compounds that will only degrade under aerobic conditions, THF, benzene and phenol, only start to disappear after the plume has exited the area covered by the interim cap. Tests have shown that small amounts of oxygen also start to appear in the area after the cap.

The only compound that seems to not follow the biodegradation theory is chloroethane. This compound is not readily degradable. However, benzene and phenol have been shown to be co-metabolites for chlorinated hydrocarbon degradation with pseudomonas bacteria.

The important point is that other factors have to be considered when determining the fate and transport of organic compounds. The Seymour site provides some compelling evidence that biological reactions play a significant role in the actual movement.

SUMMARY

From what we have learned from the analysis of retardation and the groundwater quality data from the site, we have realized that we cannot and should not rely solely on the transport model in its present form to design the remedial pumping system. The site data indicate that biochemical processes are an extremely important factor in the transport of contaminants and must be taken into consideration. In fact, they are so important that we are proposing a remedial pumping network which will allow the maximum in situ biodegradation to occur at the site. We are also attempting to add a factor to the solute transport model which will account for biochemical reactions so that we can refine predictions of cleanup times and estimates of cost for the project.

REFERENCES

1. **Hauptmann, M. G., J. Rumbaugh, and N. Valkenburg.** "Use of modeling during Superfund cleanup." in *Proc. 11th National Superfund Conf. Hazardous Materials Control Research Institute,* 1990.
2. **McDonald, M. G. and A. W. Harbaugh.** *A Modular Three-Dimensional Finite-Difference Groundwater Flow Model,* USGS TWRI Book 6, Ch. A1, 1988.
3. **Reeves, M., D. S. Ward, N. D. Johns, and R. M. Cranwell.** *Data Input Guide for SWIFT II,* Sandia National Laboratories, NUREG/CR-3162, 1986.

CHAPTER 19

Hydrogeologists Should Manage Biological In Situ Remediations

Evan K. Nyer

INTRODUCTION

One of the unusual circumstances that I have noted in the field of re-mediation is that we rely on microbiologists to manage in situ bioremediations. Although microbiologists are an important part of the project, I do not un-derstand why they should be responsible for making most of the important decisions. An in situ bioremediation project relies on combining the expertise from several disciplines in order to develop a successful project. This chapter will review the details necessary to understand the decisions that must be made during the design and operation of an in situ bioremediation project.

An in situ bioremediation project is based upon biochemical reactions occurring in a geological setting. Although bacteria are responsible for the

ISBN 0-87371-731-7

biochemical reactions, their natural degradation rate is limited by chemical and physical factors. An in situ bioremediation design requires the identification of the rate-limiting factors of the bacteria and their delivery to the bacteria.

WHAT ARE THE MAIN COMPONENTS OF BIOCHEMICAL REACTION RATES IN AN IN SITU PROJECT?

An in situ bioremediation project is made up of four major components:

- Microorganisms
- Oxygen
- Nutrients
- Environment

The microorganisms are the workhorses of the project. The bacteria use the organics that were released to the environment as a source of food. Chemical bonds in the organic molecules act as the source of energy for the endemic bacteria and as building blocks for reproduction. Bacterial growth and reproduction occur naturally, not because they comprehend the regulatory consequences of a contaminant plume.

The large amount of time and money that has been spent on in situ projects has gone toward trying to determine whether the bacteria that are necessary for the degradation are present at the site or if specialized bacteria must be imported. The most expensive approach to answer this question is to try to identify the bacterial species that can degrade a specific compound found at the site. In reality, multiple microorganisms work in concert during the degradation of an organic compound. In the field, a single bacterial specie is never responsible for site remediation.

If the compounds are degradable, then the natural bacteria at the site are usually able to degrade the compounds. The only times that bacteria need to be introduced to a site is when a toxic condition existed at the site and killed all of the natural bacteria. After the toxic conditions have been neutralized, bacteria may have to be reintroduced to the site in order to speed the cleanup. New spills of degradable material can also be toxic to large portions of the natural bacteria. The cleanup of new spills may be enhanced by introducing mixed bacterial cultures.

These circumstances assume that the organics are degradable. In general, petroleum hydrocarbons are degradable and chlorinated hydrocarbons are less degradable. The more chlorine substitutions on the organic compound, the less degradable the compound. (See Chapter 3 for a list of the degradability properties of the 50 organic compounds most often found at contaminated sites.)

There are many research projects that are evaluating the application of specialized bacteria for hard-to-degrade organic compounds. However, no one has successfully demonstrated the advantages of adding a specialized bacteria to a site and published the detailed information. The USEPA conducted a thorough study of bacterial additives for the Alaska oil spill cleanup. None of these products produced significant improvement when applied in the field. Many other research programs are ongoing for the degradation of chloridated hydrocarbons and PNAs. Even when these products have been developed, delivery of the bacteria to the contaminated zone will still limit use in an in situ bioremediation project.

While the bacteria are the key to bioremediation, at the present time we cannot affect whether the appropriate bacteria are present at the site of the organic contamination. In most cases, if the compound is degradable, the natural population has already adapted to the available organic compounds and is using the compounds as a food source. Simple microbial tests can be conducted on the soil to confirm the presence of viable bacterial populations and those that are capable ot degrading the specific organic compound that was released. A microbiologist should be used to conduct the testing of soil and aquifer samples to confirm the presence of the appropriate bacterial populations.

The real object of an in situ project is to enhance natural bacterial growth and reproduction. We do this by supplying the factor that is limiting the reaction rate of the bacteria. The main limitations are oxygen, moisture, and nutrients, NH_3, and PO_4. We must also ensure that the environment is suitable.

Oxygen is the main rate limiting factor in organic chemical degradation. The bacteria need large amounts of oxygen to produce energy. Many in situ projects have required oxygen without nutrient addition. However, none, to my knowledge, have excluded oxygen and only required nutrient addition.

Moisture is the second most important factor. Of course, moisture is only an important factor in the unsaturated zone and not the aquifer. If the unsaturated zone contains too little moisture, then the bacteria will not have the microenvironments that they need to survive. Too much moisture will inhibit oxygen transfer. Laboratory tests can be run to determine the affect of moisture content. To monitor this factor on a full-scale system, humidity measurements can be made on the effluent air stream of a VES/Bio system. This will evaluate the moisture content of the soil.

The bacteria also require macronutrients and micronutrients to reproduce. The macronutrients are nitrogen in the reduced form, NH_3, and phosphorus in the most oxidized form, PO_4. Micronutrients are almost always present in either the soil or aquifer and do not have to be considered in an in situ project. Nutrients are needed in situations where there is a need to grow large bacterial population. This would be appropriate for large spills or when it is necessary to minimize the total project time. Both macro- and micronutrients can interact with the soil matrix. Before a nutrient delivery system is designed, a geochemist should conduct a series of tests to ensure that the nutrients can migrate through the soil and aquifer.

The final factor that must be considered is the environment. Bacteria grow best under certain temperature and pH conditions. In general, the higher the temperature, the faster the bacteria will reproduce. Although aquifers maintain a relatively constant temperature during the entire year, the best time to start up an in situ project is during the warmer months. The aquifer will maintain activity all year, but the unsaturated zone may shut down biological activity in the winter months in the northern states. pH should be kept between 6 and 8.5.

All of these factors have one thing in common; they must all be delivered to the contaminated zone. As stated above, the biological reactions are probably already occurring at the site. The main challenge in an in situ design is the enhancement of the natural, ongoing reactions. Although laboratory tests can be used to determine whether oxygen, moisture, and nutrients will increase the rate of reaction, the real design problem is how to deliver these factors to the bacteria at the contaminated zone.

In order to understand why the delivery of these materials is difficult, we need to have a better understanding of the relationship of the organics to the aquifer and unsaturated zone. Initially, we must understand why in situ is an important process in the first place.

WHY DO WE USE IN SITU TREATMENT?

The best answer to this question is that the only way we can actually clean up an aquifer is by biological in situ treatment. We have previously discussed the limitation of pump-and-treat systems. Other authors have now published that pump-and-treat is not capable of remediating an aquifer. Many people have realized that while you pump to control plume movement, you actually use in situ to remediate the plume. Of course, the first time that I heard this was from Dick Raymond ten years ago.

The primary reason that we cannot use pumping systems to remediate an aquifer is that the aquifer does not release all of the contaminants at the same time. Figure 1 shows the life-cycle concentration during a remediation. At first, the contaminants quickly leave the aquifer with the pumped water. However, as the pumping continues, less and less mass of material is removed from the aquifer. The contaminant almost stops decreasing near the end of the project. The flattening of the line is described as the asymetope, and many projects try to use this flat area to define the end of the project. The problem with using the asymetope is that if the pumping stops, the concentration can increase in the aquifer. The same pattern holds for the unsaturated zone and a vapor extraction system.

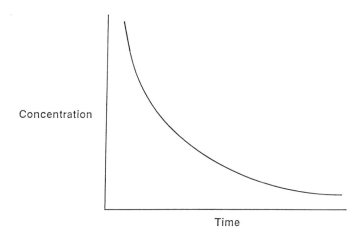

FIGURE 1. Life-cycle concentration during a remediation project.

There are several ways to describe why the concentration stops decreasing and increases if the pumping stops. One of the best visual methods comes from Szecsody.[1] Figure 2 shows the four main mechanisms that control the rate of release of organic compounds into the water within the aquifer. (I would also suggest that the reader review Keely's article[2] for its discussion on the relationship between solubility, diffusion, advection, and desorption on the concentration of contaminants in the water of an aquifer.) As can be seen in Figure 2, not all of the organic material is in contact with water that is moving. The organics in contact with stagnant (or relatively stagnant) water must overcome simple diffusion in order to be removed from the aquifer.

We can now see the relationship between Figures 1 and 2. The beginning of the project produces high concentrations in the pumped water due to the organics, in direct contact with flowing water, being quickly removed from the aquifer. The concentration decreases as those organics are removed. The end of the project has a relatively constant concentration of organic compounds due to the slow diffusion of organics from stagnant water into flowing water.

We can also explain why the concentration increases if the pumping system is turned off. When the pumps are turned off, the flowing water returns to its natural, slow movement. The organics in the stagnant water continue to diffuse into the slower moving water. The mass of organics that diffuses is the same, but the amount of water that passes through that area of the aquifer decreases. The same mass of organics in less water creates a higher concentration. The concentration increases until a new equilibrium is established between the diffusing organics and the slower moving water. When a water sample is taken from a monitoring well, the concentration reflects the new equilibrium of the aquifer.

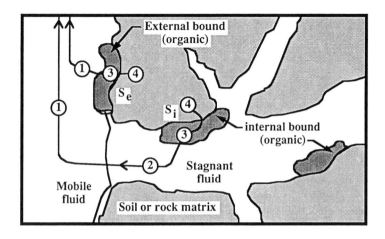

FIGURE 2. Conceptual model of sorption in porous aggregate. (1) Solute transport in the mobile fluid and stagnant boundary layer. (2) Intraparticle diffusion in stagnant fluid. (3) Transport in bound organic phase. (4) Binding and release within the bound organic phase and/or at the mineral surface.

The final point of this section is the main focus of this chapter. The key to a successful design of a pumping system is to minimize the microcosms that are in stagnant or relatively stagnant water flow conditions. A groundwater recovery design must maximize the flow across the part of the aquifer that is contaminated. The wells and recharge system must set up a pattern of flow that accomplishes this objective. The knowledge and experience of a hydrogeologist is required to meet this performance critiera.

HOW DO WE DELIVER THE NECESSARY MATERIAL TO THE BACTERIA?

Now let us turn the discussion around and, instead of removing material from the aquifer, let us look at the processes involved in delivering material to the microenvironments in the aquifer.

Figure 2 shows us where the organics are located in a contaminated aquifer. We can assume that bacteria exist through the entire microcosm. Bacteria are present in the flowing region and in the stagnant zones.

These bacteria will interact with the organics present. As stated before, they will use the organics as a food and energy source. The rate at which they use these organics is limited by oxygen, moisture (in the unsaturated zone), and nutrients content of the microcosm. In order to increase the rate of bacterial reaction and reduce the time for remediation, the rate-limiting factors must be delivered to the microcosm where the organics and the bacteria are already present.

Now we have the same problem that existed when we tried to design a removal system. We have flowing water areas and stagnant areas in the aquifer. The flowing water areas can have a direct delivery of the growth-limiting factors. However, the stagnant areas require that the oxygen and nutrients diffuse to the specific sites. Only by diffusion can we deliver the oxygen and nutrients to the site where the biological reaction is taking place in the stagnant zones.

In a removal design, minimizing the stagnant areas was important. In an in situ design, minimizing the stagnant areas is critical. In fact, the success or failure of the system is based upon the ability of the system to deliver the enhancement factors to the locations that need them. Since diffusion is a relatively slow process, we must minimize the area of the aquifer that it relies on. Nothing will affect the time for cleanup more than the delivery of the enhancement factors.

WHO SHOULD RUN AN IN SITU PROJECT?

As can be seen in this discussion, many experts are required for the proper design of an in situ remediation. Areas of expertise that are needed include:

- **Microbiologists** — to ensure that the proper bacteria are present at the site, that there are no toxic conditions at the site, and what material is required to enhance the rate of reaction of the bacteria
- **Geochemists** — to ensure that the material that must be delivered to the bacteria will not interact with the soil particles in the aquifer and unsaturated zone
- **Hydrogeologists/Geologists** — to first develop the data necessary to define the extent of contamination and, second, to determine the best way to deliver the required material to the bacteria that will enhance their rate of reaction
- **Engineers** — to design the aboveground equipment necessary to deliver the material

While all of these experts are required as part of the project, the question remains whose expertise is the most important and, therefore, who should make the final decisions during the project. The methods for delivering the required oxygen, moisture, and nutrients are the critical portion of the in situ remediation. The hydrogeologist is the expert that must develop the system that will deliver these factors to the bacteria.

Hydrogeologists are the best choice for running an in situ remediation project. They have the proper education, experience, and understanding to make the critical design and operation decisions during a remediation project. All of the failures that I have seen with in situ programs were the result of the geology or hydrogeology of a site. I have never seen the bacteria (or lack of bacteria) be the cause of an unsuccessful in situ project. If bacteria are the reason that an in situ project will not work, this fact can be discovered early in the project with simple laboratory tests.

Once again, all of these experts are required to have a successful in situ project. But the critical points of designs and operations are delivery of enhancing material. A hydrogeologist has the best chance of designing and running a successful remediation.

REFERENCES

1. **Szecsody, J. E. and Bales, R. C.** *J. Contam. Hydrogeo.* 4: 181-203 (1989).
2. **Keely, J. F.** "Performance Evaluation of Pump-and-Treat Remediations." *EPA Ground Water Issue.* EPA/540/4-89/005 (October, 1989).

CHAPTER 20

How to Determine the End of Active Remediation

Evan K. Nyer and Douglas G. Mehan

When we discuss pump-and-treat or in situ groundwater treatment systems, we rarely spend a lot of time discussing the end of the project. A tremendous amount of time, effort, and money are spent on defining the beginning of the project. The site assessment can take years and cost anywhere from $50,000 to $5,000,000 or more. No cost is too great to determine the concentrations at the start of the project. We will take these concentrations and spend more money to perform a feasibility study to determine the optimum design for the treatment system. All of this is necessary (well, maybe, not all of it) to develop the correct solution to the original contamination. However, we do not spend a relative amount to determine the end of the project.

This does not mean that we completely ignore the end. Most remediation plans include a final concentration that the project is trying to achieve.

ISBN 0-87371-731-7

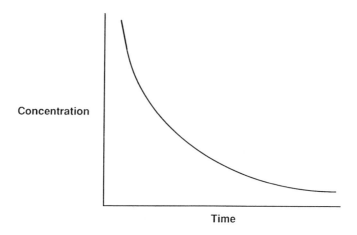

FIGURE 1. Life-cycle concentration during a remediation project.

However, we do not analyze what these concentrations mean to the design of the treatment system and the operation of the project in general. This chapter will review how we define the end of the project and how to establish a strategy to reach that finish point.

LIFE-CYCLE OF A PROJECT

Figure 1 shows the normal life-cycle concentration of a remediation project. Normally, we concentrate our discussions on the beginning portion of the curve. We cannot simply use the concentration that we find during the remedial investigation as the only criteria to design a groundwater treatment system. The concentration changes over the life of the project and any design must address all of the concentrations encountered during the entire project.

Figure 1 is a conceptual plot of concentration vs time. The plot shows that, as time increases, the concentrations decrease in a nonlinear relationship. This chapter will concentrate on the bottom part of the curve where concentrations change at very slow rates. As the project progresses (time), the rate of removal (concentration) decreases. Figure 1 shows the curve becoming almost parallel to the horizontal axis over a period of time. There are several processes that may contribute to the flattening of this curve, all of them may, in part, affect water quality. These processes include physical processes (dilution, dispersion, filtration, and gas bubbles), chemical processes (complexation, acid-base reactions, redox reactions, precipitation-dissolution, and sorption-desorption), and chemical reactions (decay, respiration, degradation, and co-metabolism).

WHAT IS CLEAN?

The objective of most remediations is to clean the site. But to meet these objectives we must ask two questions. What is clean and how do we define it? The three main conservative methods used to establish clean are risk assessments, federal drinking water standards, and analytical detection limits. All of these methods have advantages and disadvantages.

Risk assessment can be uniquely designed for the specific site. Any unusual paths for human contact, and any design method used to prevent human contact can be incorporated into the risk assessment. However, there are no official standard methods to develop the risk formulations. Different basic assumptions will result in different specific numbers calculated for clean. Some states will not allow risk assessment to be used due to the possible variability of the results from the data. While risk assessment may be the only method that can consider local anomalies, the variable output may cause long discussions on the reality of the numbers.

The federal drinking water standards are another source of numbers that can be used to establish what is clean. These numbers are set and specific. Their basis have been published and discussed before the final figures were made official. Over time, more compounds will be included on the list. The only problem for remediation sites is to establish the relationship between the soil and the ground water. All of the drinking water standards are based only on the concentrations in water. Problems may occur when organics sorbed (absorbed or adsorbed) to the soil particles in the unsaturated zone and the aquifer may release more slowly to the ground water. While drinking water standards have a strong technical basis, this method does not directly address the sorbed material.

Analytical detection limits are the third method for establishing what is clean. For compounds that are highly toxic or in situations in which the numbers developed during a risk assessment are less than the detection limit of the compound, the analytical detection limit can be used to establish what is clean. The main problem with this method is that the ability to detect each specific compound continuously improves and these detection limits decrease. No one should accept analytical detection limits as clean. Instead, the detection limits should be used as a basis for a specific concentration. Without a specific number, what is clean may change over the life of the project.

Any of these methods can be used to determine what is clean. However, if we look at each of these methods, we are basing these concentrations on contact with human beings. In other words, clean is when the ground water or soil is safe for human consumption. The end of the remediation is when the site is safe for people.

The problem with this definition of clean is shown in Figure 2. As the site gets closer to clean, the contaminant concentration reaches its asymetote.

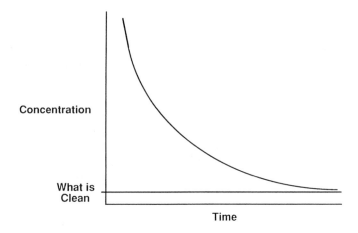

FIGURE 2. Achieving "clean" during a remediation project.

This figure represents a worst case scenario in which the site never reaches clean. Even in cases where the site does reach clean, it will take many years.

During the last years of the project, the treatment system suffers from diminishing returns. The treatment system continues to run, but the amount of material removed is minimal. The money is still being spent on the site, but there is little return for this financial outlay.

RETARDATION VERSUS BIOCHEMICAL ACTIVITY

A second factor comes into consideration when we approach the asymetote of the life-cycle concentration. Natural biochemical reactions may be occurring at the same rate as the natural release of the contaminants due to retardation factors.

As we have discussed in several previous chapters, natural bacteria exist throughout the soil, unsaturated zone, and the aquifer. If toxic conditions do not exist, then the natural bacteria are probably already degrading the contaminants. In general, petroleum hydrocarbons degrade quickly and chlorinated hydrocarbons degrade slowly. The bacterial rate of degradation is also limited by the presence of oxygen (usually the main limiting factor for degradable compounds) and nutrients. In fact, biodegradation models, like BIO-PLUME II,[1] use oxygen concentration as the limiting factor in degradation rate calculations.

The natural rate of degradation is small compared to an aboveground reactor or enhanced in situ biological reactions. The aquifer or unsaturated zone have a limited capacity to replenish oxygen used by the bacteria. But, when the contaminant concentration reaches very low levels, the aquifer and unsaturated zone can naturally supply what is needed.

At the same time, all of the contaminants do not release to the water at the same rate. The contaminants are sorbed to the soil or aquifer. Depending on the organic content of the soil and the chemical properties of the contaminant, the individual compounds may release to the water at different rates. Several contaminant transport processes may cause residual contamination to be flushed from the aquifer more slowly. Transport processes that may cause this include: diffusion or contaminants within heterogeneous sediments, hydrodynamic isolation, sorption-desorption, and liquid-liquid portioning.[2]

In the previous chapter of this section we used these same arguments to demonstrate that a plume of degradable organics would not move through an aerobic section of an aquifer. This data was from an actual site, and we were able to show that pumping would not improve the remediation. In fact, a higher mass of compounds would be exposed to humans from a pumping system than by allowing the natural bacteria to degrade the compounds in situ. The previous chapter also demonstrated that natural degradation could occur at the same rate as plume migration. This chapter is based on the same technical points, but takes them one step further to a complete elimination of the plume.

At some stage in the life cycle, the rate that the compounds can be removed from the aquifer by pumping will be the same as the rate of natural biological degradation in the aquifer. And, we will not be able to speed the cleanup no matter how fast we pump the well. In fact, if the aquifer can naturally replenish the oxygen demand (and other minor rate-limiting factors) from the contaminants, then pumping will not increase the rate that we reach clean.

Once we reach this point, then, all of the money that we are spending on pumping and treating this water is going to waste. The pumps could be turned off and all of the equipment removed, and we would still reach clean in the same amount of time as leaving the system running.

ACTIVE MANAGEMENT

We need to develop another level in the life cycle of the project that determines when we can turn off the treatment system. This is the stage where we can stop spending a majority of the money. Clean is where we can reuse the property. Active management is when we can stop spending money.

Figure 3 shows the life-cycle curve with the clean line and an active management line. The active management line represents the stage in the life cycle where pumping will no longer speed the cleanup of the site. We should stop spending money at this time. The clean line represents when we can use the water and the site for human consumption.

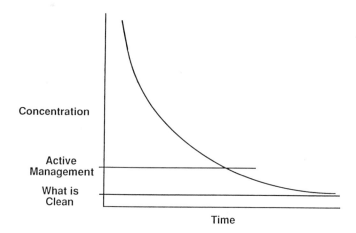

FIGURE 3. Active management vs what is clean during a remediation project.

The situation is similar to the 1970s when we cleaned up the rivers and lakes (well, started to). We installed wastewater treatment systems on municipal and industrial wastewater facilities. That stopped the contaminants from entering the water body (lake or river). The river or lake then cleaned itself. This took time, and we did not use the water body the day after we installed the treatment system. We waited until the water was clean or safe to use.

Rivers and lakes can remediate themselves faster than ground water. One reason for this is that oxygen can transfer into surface water bodies faster than into ground water. But, ground water can remediate itself. The problem is that the rate is slow and we normally find the timeframe unacceptable. However, if we remove most of the contaminants, then the aquifer can finish the job on its own. If this removal rate is the same as remediation by pumping and treating, then there is no reason to continue the pumping and treating. We are at the point in the life cycle of the project that we must wait before we can use the water. Active management of the site should stop, and we should only monitor the site while we wait for clean.

BIO MODELLING

The last problem is to determine when we reach active management in the life cycle of the project. This point is basically a comparison between the normal fate and transport of the chemicals and the biochemical reaction rates of those same chemicals. Computer modelling will then be used with field and laboratory studies to help predict when the active management point is reached.

The solute transport models will need to be expanded to include biochemical reactions. The mathematics of the solute transport computer model will need to be changed to include a disappearance of a compound and not simply a retardation. The rate of this disappearance will be dependent upon the compound (degradability), bacteria (sufficient numbers present), and the aquifer (natural oxygen transfer and nutrient availability).

We originally used computer models to help us design a capture system for the contamination plume. We now need to expand the models in order to determine the end of the active management of the remediation project.

REFERENCES

1. **Rifai, H., et al.** "Biodegradation Modeling at Aviation Fuel Spill Site", *J. Environ. Eng.* 114 (5) (October, 1988).
2. **Johnson, R. L., et al.** "Transport and Fate of Contaminants in the Subsurface", U.S. Environmental Protection Agency Seminar Publication, EPA/625/4-89/019, 148 (1989).

Contributors

Paul Bitter, P. E. is an environmental engineer for Geraghty & Miller, Inc., involved with all facets of treatment of hazardous waste sites with an emphasis on bioremediation of ground water and sludges. He has worked mainly on projects that involve the bench and pilot testing of UV/oxidation, bioremediation, and steam-stripping systems for the treatment of organic constituents found in ground water.

Gary Boettcher is a project scientist with Geraghty & Miller, Inc. in Tampa, Florida. He received his B.S. degree in microbiology from the University of South Florida and is currently pursuing a Master of Public Health (MPH) degree. He is involved in investigations, treatability, and design of biological remediation systems.

Peter Boutros has 18 years experience in engineering, design, and maintenance of electrical and control systems. He is a member of NFPA and a senior member of ISA.

David M. Keyser has over 13 years' experience in manufacturing, process, and plant engineering. With three years at Geraghty & Miller, he is presently a Staff Mechanical Engineer and Project Manager overseeing designs of projects in groundwater and soils remediation. His other responsibilities include consulting in plant operations' design and field support.

Victoria Kramer acts as project scientist for Geraghty & Miller and is involved with several CERCLA-, RCRA-, and ECRA-mandated investigations. Kramer is the assistant project manager for the Seymour, Indiana project and is responsible for preparing work plans and scheduling field and support personnel. Kramer is a graduate of C.W. Post University and holds a B.S. in environmental geology. She is working on a master's degree at the State University of New York at Stony Brook.

Douglas G. Mehan, P. G., is a hydrogeologist with Geraghty & Miller, Inc., Tampa, Florida, where he serves as an environmental and water supply project manager, responsible for hydrogeological investigations. His professional interests lie in the areas of contaminant fate and transport.

Jodie Montgomery is a specialist in the completion of feasibility and corrective measure studies for remediation of ground water, air, and soils at CERCLA and RCRA sites. She has worked with numerous regulatory agencies in the negotiation of consent decrees for hazardous waste site management.

Bridget Morello received a B.S. from the University of South Florida in 1987 and is currently working for Geraghty & Miller's Process Group in Tampa, Florida. She is involved mainly in treatability evaluation and design of groundwater treatment systems.

Gregory J. Rorech, P. E., is the Office Manager for Geraghty & Miller, Inc.'s Jacksonville office. He specializes in the design of recovery, treatment, and discharge systems for contaminated aquifers. He has conducted remedial evaluations at several CERLA sites as well as numerous other hazardous waste sites. He obtained a B.S. in chemical engineering from S.U.N.Y. at Buffalo.

George J. Skladany is Manager of Technology Applications for Envirogen, Inc., an environmental biotechnology company located in Lawrenceville, New Jersey. His general responsibilities include evaluating and recommending the suitability of using biological and physical/chemical treatment technologies in above-ground and in situ remediation systems. Specific responsibilities include providing site remediation and pollution prevention implementation direction for aqueous waste and air toxics problems. Skladany previously served as Manager of the Engineering Treatability Lab for General Physics Environmental Services (Gaithersburg, MD); Manager of Operations for DETOX, Inc., (Dayton, OH); Field Microbiologist for O.H. Materials Co. (Findlay, OH); and On-Sceen Coordinator with the US EPA Superfund Program (Lexington, MA). Skladany holds a B.S. degree in Cellular and Molecular Biology from Carnegie-Mellon University (1978) and an M.S. degree in Environmental Systems Engineering from Clemson University (1984). Over the past six years, he has instructed classes in bioremediation and ground water treatment for the US EPA, the UCLA Extension Program in Toxic and Hazardous Materials Management and Control, the American Society for Microbiology and other professional organizations.

Nicholas Valkenburg manages Geraghty & Miller's Plainview, New York office and is also responsible for several large site investigations and remedial projects. He specializes in CERCLA- and RCRA-mandated investigation and remedial programs and is the project officer for the Seymour, Indiana project which involves soil bioremediation, the installation of a RCRA cap, a vapor extraction system, and a groundwater pumping and treatment system. Valkenburg holds a B.S. in geology from the State University of New York at Stony Brook and an M.S. in geology from the University of Toledo. He was also an adjunct faculty member of Adelphi University where he taught courses in hydrogeology and groundwater management.

Index